New Directions in Philosophy and Cognitive Science

Series Editor
Michelle Maiese
Emmanuel College
Boston, MA, USA

This series brings together work that takes cognitive science in new directions. For many years, philosophical contributions to the field of cognitive science came primarily from theorists with commitments to physical reductionism, neurocentrism, and a representationalist model of the mind. However, over the last two decades, a rich literature that challenges these traditional views has emerged. According to so-called '4E' approaches, the mind is embodied, embedded, enactive, and extended. Cognition, emotion, and consciousness are not best understood as comprised of brain-bound representational mechanisms, but rather as dynamic, embodied, action-oriented processes that sometimes extend beyond the human body. Such work often draws from phenomenology and dynamic systems theory to rethink the nature of cognition, characterizing it in terms of the embodied activity of an affectively attuned organism embedded in its social world. In recent years, theorists have begun to utilize 4E approaches to investigate questions in philosophy of psychiatry, moral psychology, ethics, and political philosophy. To foster this growing interest in rethinking traditional philosophical notions of cognition using phenomenology, dynamic systems theory, and 4E approaches, we dedicate this series to "New Directions in Philosophy and Cognitive Science."

If you are interested in the series or wish to submit a proposal, please contact Amy Invernizzi, amy.invernizzi@palgrave-usa.com.

Paulo Alexandre e Castro
Editor

Challenges of the Technological Mind

Between Philosophy and Technology

Editor
Paulo Alexandre e Castro
Institute for Philosophical Studies
University of Coimbra
Coimbra, Portugal

ISSN 2946-2959   ISSN 2946-2967   (electronic)
New Directions in Philosophy and Cognitive Science
ISBN 978-3-031-55332-5   ISBN 978-3-031-55333-2   (eBook)
https://doi.org/10.1007/978-3-031-55333-2

Cover illustration: © duncan1890/Getty images

This Palgrave Macmillan imprint is published by the registered company Springer Nature Switzerland AG.
The registered company address is: Gewerbestrasse 11, 6330 Cham, Switzerland

Paper in this product is recyclable.

Dedication
To Alice and Laura
always in my (technological) mind

ACKNOWLEDGMENTS

Anyone who has ever edited a book knows that between having the idea and turning it into a book dozens of things happen. If, on the one hand, this makes the experience internally challenging (although often stressful), on the other hand, externally it becomes rewarding and enriching; not just because the book is finished, but because one of the most enriching aspects of the entire process is seeing and feeling the commitment and joy of everyone involved.

I would like to begin by thanking all the contributors (not only authors) to this book for their wisdom and perseverance but also their generosity and commitment; I especially thank the authors who helped make this happen: Adelaide Costa, Ania Malinowska, Alessio Plebe, David Rosenthal, Juraj Hvorecký, Lucia Santaella, Marcin Miłkowski, Patrícia Gouveia, Pietro Perconti and Ralph Ings Bannell. I must also express my gratitude to all the people involved in the editing team of Palgrave for their patience, professionalism and generosity, especially Amy Invernizzi and Antony Sami. This book exists because of you all.

I should also add that working with all of you gave me the opportunity to continue to believe that the world is a better place thanks to people who want to evolve and help others to evolve, academically, professionally, humanly. So, what makes all of this even better, more rewarding (if I may say so), is that these are people who selflessly share the gift of their time, their knowledge, their wisdom.

CONTENTS

NOTES ON CONTRIBUTORS

Paulo Alexandre e Castro is a full member of the Institute for Philosophical Studies at the University of Coimbra (IEF-UC, Portugal) and visiting Professor at the University of Caxias do Sul (Brazil). He has taught at the University of Lisbon and University of Minho (among others institutions like Polythecnic University of Viseu). He has a PhD in Philosophy of Mind (University of Minho), a Master's in Phenomenology and Hermeneutics and a degree in Philosophy from the Faculty of Arts and Humanities (University of Lisbon). He has a Post-doctorate in Digital Art/Cyberliterature (Fernando Pessoa University), as well as completing other courses, such as "Ethics in Artificial Intelligence" (University of Helsinki-Finland) or "Humanidades Digitales" (Universidade de Barcelona-Spain). He has a PDCHyp (from the London College of Clinical Hypnosis) and an author of 13 books (7 essay collections and 6 works of literature) and also the co-author of more than fifty books. He also publishes regularly in international journals. His research interests are in the areas of neuroaesthetics, philosophy of art, philosophy of mind, education and contemporary culture and art. More information about the author, consult Philpeople, ResearchGate, Academia.edu or Orcid.

Ralph Ings Bannell is a (retired) Professor of the Philosophy of Education, at the Pontifical Catholic University of Rio de Janeiro. He has a degree in Philosophy (first class) from Stirling University and an MA and DPhil. in Social and Political Thought from the University of Sussex. He is a founding member and past President of the Brazilian Society for the Philosophy of Education. His interests lie mainly within two areas: (1) the interface between social and political philosophy and education, exploring topics such as equity, equality, democracy, freedom and citizenship; and (2) the interface between the philosophy of cognition and mind and education, exploring topics such as e-cognition, especially embodied and enactive theories, embodied learning, developmental, cognitive and ecological psychology and learning theories. He also has a strong interest in epistemology, ethics and the social, political and cultural aspects of second-language teaching and acquisition, as well as "intelligent technologies" and education.

Adelaide Costa, MD, PhD in Bioethics, in the area of Neuroenhancement, by the Portuguese Catholic University. Her thesis was awarded in recognition of its contribution to Bioethics. She is a psychiatrist and Professor of Psychiatry at the Department of Clinical Neurosciences and Mental Health, Faculty of Medicine, University of Porto. She is the author and co-author of several articles published in national and international journals and also of several lectures, workshops and courses in the areas of Psychiatry and Neuroethics, presented in both congresses and universities. She has collaborated with the Institute of Research and Innovation in Health at the University of Porto (i3S) and also in in teaching with the Faculty of Medicine of the University of Coimbra and also the Portuguese Catholic University. She currently works as an expert in forensic psychiatry at the Institute of Legal Medicine. Her main areas of clinical work include pain and palliative care, psychosomatics, gerontopsychiatry and forensic psychiatry and her main areas of research are in neuroenhancement, forensic psychiatry, humanization in healthcare and the psychosocial impact of the COVID-19 pandemic.

Patrícia Gouveia is Associate Professor at Lisbon University Fine Arts Faculty (Faculdade de Belas-Artes da Universidade de Lisboa, FBAUL) and also an integrated member of ITI—Interactive Technologies Institute/LARSyS, Laboratory for Robotics and Engineering Systems, IST. He is the co-curator of the *Playmode* exhibition (MAAT 2016–2019) and has worked in multimedia arts and design since the 1990s. Her research focuses on playable media, interactive fiction and digital arts as a place of convergence between cinema, music, games, arts and design. For more information about the author and her works please see here: https://fbaul.academia.edu/PatriciaGouveia/CurriculumVitae.

Juraj Hvorecký is a researcher at the Department of Applied Philosophy and Ethics, as well as at the Center for Environmental and Technology Ethics – Prague, at the Institute of Philosophy of the Czech Academy of Sciences. He also teaches at the Undergraduate Program in Central European Studies (UPCES) program in the Czech capital. He combines his interests in philosophy of mind with applied ethics, especially in the domain of disruptive technologies. More specifically, he is interested in emotions, conscious/unconscious divide, and the ethical implications of social robotics and autonomous technology. Juraj is a member of the Czech National Bioethics Committee and the Czech Ethics Committee for Autonomous Mobility. He is among the coeditors of *Conscious and Unconscious Mentality*, to appear in 2024.

Ania Malinowska is an author, a cultural theorist, Associate Professor in Media and Cultural Studies at the Faculty of Humanities, University of Silesia (Poland), and a former Senior Fulbright Fellow at the New School in New York. Her research concentrates on cultural theory, love studies, digital humanities and critical robotics—and specifically on the formation of cultural norms and the social, emotional and aesthetic codes that relate to digitalism.

Malinowska has authored and co-edited a number of articles, chapters and books preoccupied with the posthuman condition and technologies of affect, including *Love in Contemporary Technoculture* (2022), with Valentina Peri), *Data, Dating. Love, Technology, Desire* (2021, with Michael Gratzke), *The Materiality of Love. Essays on Affection and Cultural Practice* (Routledge, 2018, with Karolina Lebek), *Materiality and Popular Culture. The Popular Life of Things* (Routledge, 2017, with Toby Miller) and "Media and Emotions: The New Frontiers of Affect in *Digital Culture*" (a special issue of *Open Cultural Studies*, 2017).

Marcin Miłkowski is Associate Professor in the Section for Logic and Cognitive Science at the Institute of Philosophy and Sociology, Polish Academy of Sciences. He published *Explaining the Computational Mind* (2013), and was awarded the Tadeusz Kotarbiński Prize of Section I of the Polish Academy of Sciences and the National Science Center Award for outstanding young scholars in social sciences and humanities in 2014. He was presented with Herbert A. Simon Award by the Association for Computers in Philosophy (IACAP) for his significant contributions in the foundations of computational neuroscience (2015). He is the chair of the

Section of Logic and Cognitive Science at the Institute of Philosophy and Sociology, Polish Academy of Sciences (2020–) and also the Deputy President of the Committee for Philosophical Sciences (2020–) of the Polish Academy of Sciences. With R. Poczobut, he edited the volume *Analytic Metaphysics of Mind* (in Polish, Warszawa 2008) and *Companion to the Philosophy of Mind* (in Polish, Kraków 2012); with K. Talmont-Kamiński, he has published *Beyond Description. Naturalism and Normativity* (2010) and *Regarding the Mind, Naturally: Naturalist Approaches to the Sciences of the Mental* (2013). The scientific interests of Prof. Miłkowski focus on philosophy of science, including philosophy of cognitive science, and philosophy of mind and information. He is also interested in computational linguistics.

Pietro Perconti is currently the Dean of the Department of Cognitive Science (COSPECS), and formerly Deputy Vice-Chancellor in charge of Teaching and Education, Director of the Master's in Cognitive Science, and Member of the Boards of Directors at the University of Messina. His current research interests include social cognition, consciousness, and the social role of cognitive science. He is the author of over 100 publications, including eight books. The latest, *The Future of the Artificial Mind* (with A. Plebe) (CRC, Taylor and Francis, London, UK) is an updated picture of what Artificial Intelligence is becoming nowadays in the light of the new perspective of deep learning.

Alessio Plebe is Professor in the Philosophy of Science at the Department of Cognitive Science of the University of Messina. His main research field is neural computation, its epistemology, and its explanatory power for several cognitive functions. He has developed realistic neural models of visual object recognition, early language acquisition and moral behaviour. Currently, Plebe investigates the recent rise of deep learning, its causes and its impact on philosophy and cognitive science.

David Rosenthal is Professor of Philosophy and Coordinator of the Concentration in Cognitive Science at the Graduate Center, City University of New York, with appointments also in linguistics and cognitive neuroscience. He works mainly in philosophy of mind, philosophy of psychology, and cognitive science. He has pioneered the higher-order-thought theory of consciousness. According to that view, conscious psychological states differ from those that aren't conscious by being accompanied by a thought that one is in the target state. He has also developed the quality-space theory of mental qualitative character, according to which qualitative mental properties bear relations to one another homomorphic to the perceptible differences among a range of perceptible stimulus properties. Relatedly, he has written about the self, the unity of consciousness, the relation of thought and speech, free will, introspection, the function of consciousness, Freud, olfactory perception, the emotions, Descartes, perceptual confidence, interpretativism, and the mind–body problem. He has also published on the significance of the history of philosophy for current work in philosophy.

Lucia Santaella Is a researcher 1A at CNPq, with a PhD in Literary Theory (PUC-SP, 1973) and in Communication Sciences (ECA/USP, 1993). She is a full professor at the PhD in Communication and Semiotics and in Technologies of Intelligence and Digital Design (PUC-SP) and Director of the latter. Santaella has been an invited professor at several universities in Europe and Latin America. She published 53 books and organized 29, in addition to publishing almost 500 articles in Brazil and abroad. She received the Jabuti awards (2002, 2009, 2011, 2014), the Sergio Motta award (2005) and the Luiz Beltrão award (2010). Her last book is from 2021 and as the title of *Hiper-Híbrids Humans* (Portuguese edition, Humanos Hiper-Híbridos, Paulus ed.).

LIST OF FIGURES

LIST OF IMAGES

PART I

Echoes of Technological Mind

CHAPTER 1

Introduction

Paulo Alexandre e Castro

We are living through an exciting period in human history. Technology and knowledge have never been more available than they are now. This essentially means two things: the first one is that we have never been so close to technology and (in a sense) so close to knowledge, and the second one, is that so much progress has never been made in such a short period of time. Naturally, there are both good and less good things happening.

Technology profoundly affects our daily lives. It has become an integral part of our daily lives, of our routines, not only in terms of technological devices that facilitate the performance of our tasks (whether at a professional or leisure level) like the use of virtual cards, barcodes, virtual purchases and payments etc., but also at a medical and clinical level. If it is true that when one thinks about technology we think mainly about computers, software, GPS or smartphones, it is no less true that one can thinks about how technology has become an extension of humans themselves. We are referring not only to pacemakers, dental implants or orthopedic prostheses (or even devices that allow diagnosis, from X-rays to PET) but also to bionic eyes, chip implants, robotic exoskeletons, monitoring applications

P. Alexandre e Castro (✉)
Institute for Philosophical Studies, University of Coimbra, Coimbra, Portugal

© The Author(s), under exclusive license to Springer Nature Switzerland AG 2024
P. Alexandre e Castro (ed.), *Challenges of the Technological Mind*,
New Directions in Philosophy and Cognitive Science,
https://doi.org/10.1007/978-3-031-55333-2_1

(from clinical to sports), among many others such as different neural interface devices. Perhaps because of all this, we feel that technology is both special and fascinating, that it, is intense and impactful, and intriguing and mesmerizing. We let it enter peacefully into our lives and now we can no longer exist—or at least live with the same perspective—, without it. In fact, technology has the power to enrich or impoverish us (some studies specify a decrease in the use of memory—the google effect also known as digital amnesia—, and a decrease in the level of language and cognition as a result of digital transformation and the adaptation of everyday habits). For better or worse, nothing has had more impact on the character of our existence (in the last decades) than technology.

The scenario in which we live—we can all agree—, has changed radically due to the full advent of Artificial Intelligence, especially in the last decade. The opening, or, if one prefers, the "democratization" of AI (from algorithms in e-commerce to platforms like ChatGPT), has brought and will also bring new questions, insecurities, challenges, which require appropriate and rapid responses, as the paradigm is changing quickly and drastically (at the social, educational, economic and political levels). This change is not a prediction: it is already happening. Hard concepts like intelligence, consciousness, creativity, mind—taken as fundamentally human characteristics for centuries—are suddenly assigned, if not even transplanted into virtual, augmented or robotic objects, devices, worlds or existences. In this sense, it seems that the territory, the border of the human, has been—and will necessarily be—expanded. Therefore, it can be said that the possibility of incorporating artificial minds into our lives will be as valid as any other possibility. All of these possibilities are shaking the foundations of the human world: what will global Artificial Intelligence become? What is human creativity or imagination for? What will be the art of the future? Can social robots take on a greater role in human life? Should humans be neuro or artificially enhanced? Basically, what is the role of humans in an increasingly technological, virtual, artificial world? If, on the one hand, we feel that there is something frightening here, on the other hand, we do not see how to abdicate, ignore or dismiss such possibilities of integration into the human world.

In fact, we can't help but think that we are technological beings. We always have been. But perhaps never before have we created something that could surpass us. We would need a Nietzsche capable of thinking about this type of technological superhuman—unfortunately, there has not yet been time to think about in a more serious way (or we would at least need works that were not understood as popular literature). Much of what happens in the chapters of this book is based on some of these disturbing, but also fascinating questions.

One of these questions is the one with which this book opens Part I and Lucia Santaella was able to pose it clearly: can we say that Artificial Intelligence is intelligent? The author begins by considering what is the domain of intelligence and how it can be understood beyond the boundaries of the human (taking the contribution of cognitive sciences to this end). Taking this into consideration, learning has become the fundamental key to "understanding intelligence as a focus of comparison between human intelligence and machine learning algorithmic intelligence." Lucia Santaella's hypothesis suggests that AI systems can continue to amplify human intelligence, but not replace it, thus leaving room to ask what it means to be an intelligent human being.

In the following chapter, called "Anthropomorphizing and trusting social robots", Pietro Perconti and Alessio Plebe explore the challenges posed by the proliferation of robots in society (especially humanoid robots). Taking cognitive constraints, ergonomic concerns and intercultural issues as topics of approach, they seek to highlight the emergence of a certain (artificial) morality that must exist in human–robot interactions. It also investigates the distinctions between physical and intentional trust, highlighting the role of deference, both biologically grounded and epistemic, in human–robot relationships (fundamental to understand this brave new world).

In Chap. 4, I take into account the phenomenon of neurohacking (the conditions under which it occurs, with what sort of technology and what the main consequences are) to then address two fundamental topics in the philosophy of mind: the extended mind theory (instead of the computational theory of mind) and mind uploading. From the general confrontation between these two things and assuming some premises taken from the analysis of neurohacking, I present a new theoretical formulation regarding the technological mind that I called the Artificial Extended Mind theory (AEMt).

In Chap. 5, Adelaide Costa presents us with a reflection on neuroenhancement from a bioethical perspective. However, considering her experience as a doctor, she offers us a clinical vignette to carry out this reflection. The theoretical background serves not so much to consider the advantages or disadvantages of enhancement but more to consider the possible dichotomies that exist in the use of the concept such as treatment versus enhancement, efficacy versus safety; in short, the value of human action. The author also refers to the importance of a broader dialogue (especially outside the academic and scientific sphere), since the fields of biotechnology and bioethics are increasingly present in human social life.

The second part of the book opens with David Rosenthal's chapter called "Consciousness, theory, and mental appearance," in which he contrasts one-factor views of consciousness with two-factor views. For the first he uses Fred Dretske, Thomas Nagel and Ned Block, and for the second (two-factor views as higher-order theory of consciousness), where he himself is included, he also considers the work of Bernard Baars, Stanislas Dehanene and Lionel Naccache. From the analysis, David Rosenthal argues that one-factor views preclude a useful explanation of consciousness, but also any kind of informative description of what it means to be conscious for a mental state.

The next chapter, called "Theoretical virtues of cognitive extension", Marcin Milkowski and Juraj Hvorecký, argues that the extended mind approach to cognition can be distinguished from its alternatives, such as embedded cognition and distributed cognition, in terms of not only metaphysics, but also epistemology. According to the authors, the extended mind approach is evaluated in terms of its theoretical virtues: empirical adequacy and ideal desiderata for scientific theories. However, in its general simplicity, it raises some problems and that is what this chapter deals with comparatively.

In the same line of thought brought by Lucia Santaella—even if the starting premises are different—Ania Malinowoska's chapter investigates the idea of the artificial mind as a condition distinct from the thought categories of the human mind. To this end, the author describes "the Hypnotic AI experiment—an artistic installation in which the user interacts with an intelligent self-learning system through hypnotic induction." Thus, it verifies the "latent" or "post-material" depth of the AI that emerges from this hypnotic cycle of the AI user. According to Ania Malinowoska, the use of this experiment/environment "encourages unconstrained experimentation with human–AI interactions to expand our consciousness and overcome our prejudice".

In Chap. 9, Patrícia Gouveia begins a self-analysis, based on two artistic installations (*Digital River*, VV.AA. 1997, and *Playmode* exhibition in Portugal, 2019; Brazil, 2022) with the aim of investigating how different interaction and internet technologies influence the formation of reality around us (from playful artistic to contemporary cultural). But there is another important reflection that Patrícia Gouveia intends to show: interaction and participation are key concepts for understanding the formation of contemporary human perception but, most importantly, the ability to enhance resistance. And resistance, "it is generated not in reaction but rather in relation to contemporary circumstances and developing sciences and technologies."

In the last chapter, Ralp Ings Bannell departs from the trilogy of mind, corporeality and technology to argue that artificial systems cannot produce meaning or have experiences because they are not embodied systems. After a brief look at today's dominant theories of mind, Ralph Ings Bannell, through the lenses of John Dewey and Mark Johnson, analyzes the concepts of experience, aesthetic sensitivity and production of meaning to show why artificial systems and technologies "intelligent" are not capable of such performances/characteristics, that is, they are not subjects of experience.

Considering all the chapters presented in this book, we realize that the challenges facing the technological mind have only just begun to emerge. Important questions were raised and the authors sought to make their generous contribution with rigorous analysis, up-to-date research, critical sense and, above all, wisdom. The wisdom of those who know that questions in philosophy are more important than answers and that the mind we seek to understand is a technological mind that materially, virtually or artificially challenges the construction and transformation of the reality in which we live.

CHAPTER 2

Is Artificial Intelligence Intelligent?

Lucia Santaella

1 INTRODUCTION

This article is a synthesis of a book I recently finished under the same title: Is Artificial Intelligence (AI) intelligent? After an overview of the development of AI, chapters were developed on consciousness and intelligence in the context of the cognitive sciences. Then, learning was taken as a key to understanding intelligence as a focus of comparison between human intelligence and machine learning algorithmic intelligence.

Simply defined, AI refers to the search for the construction of machines capable of performing cognitive tasks similar to those that until then were exclusively human. To find answers to the question of whether AI is intelligent, some precautions need to be taken. First of all, peremptory claims that Artificial Intelligence is neither artificial nor intelligent should be avoided. Without failing to draw attention to the costs, risks and need for regulation of AI, statements of this type, quite sensationalist, also run the risk of propagating and not helping at all the reflections that must be carefully developed on the issue.

L. Santaella (✉)
São Paulo Catholic University, São Paulo, Brazil
e-mail: lbraga@pucsp.br

9

P. Alexandre e Castro (ed.), *Challenges of the Technological Mind*, New Directions in Philosophy and Cognitive Science, https://doi.org/10.1007/978-3-031-55333-2_2

Unfortunately, the literature on AI is full of texts that seek, in one way or another, to deny any type of intelligence to computational systems. Such denials do not help to understand what, in fact, machines are capable of accomplishing today and, conversely, they also do not clarify the differences between biological and Artificial Intelligence. Most of the time, the denial of any type of intelligence to algorithms comes from a nebulous confusion between consciousness and intelligence or the cognitive mind. Much has already been written and is already known about the latter. Consciousness, on the contrary, has so far constituted a "hard problem" for philosophers, psychologists, neuroscientists, among others (Chalmers 1995).

Without wanting to solve the hard problem of conscience, to begin with, it is not to be confused with intelligence, although, in the human animal, both are intertwined. Without getting into labyrinthine philosophical discussions, some findings from what Ford (2018) calls "intelligence architects" are quite significant. For Tenenbaum (2018: 482–483), consciousness is difficult to define because it means different things to different people. Even among cognitive scientists, philosophers, and neuroscientists, there is no consensus on this.

To simplify the author raises two aspects: in philosophy, consciousness is often referred to as qualia, that is, "the subjective sense of an experience difficult to capture in any kind of formal systems": for example, the redness of red as it is felt for each of us. "We take it for granted that the other person is seeing the same color, but we don't know if they have the same subjective experience that I do." The second aspect refers to the sense of self. We experience the world "in a certain kind of unitary way and we experience ourselves as being in it." We don't experience it in terms of tens of millions of neurons lit. Although the competence on this topic comes from psychoanalysis, in its general sense, the self comes from "I am here and I am not just my body."

Bostrom (2018: 11), on the other hand, declares that everything depends on the sense in which awareness is taken. One of these senses is the ability to have a functional form for self-awareness, that is, you are able to model yourself as an actor in the world and reflect on how different things are able to lead you to achieve this. Above all, "you think of yourself as persisting in time."

Another sense refers "to this experiential phenomenal field that we think has moral significance." For him, consciousness is a side effect of human intelligence and, thinking about artificial forms of intelligence, there is still no clear vision of what is necessary and what are the sufficient conditions for the development of morally relevant forms of consciousness. Nevertheless, the possibility that machinic intelligences can attain consciousness must be accepted.

Lecun (2018: 131) agrees with Bostron in the consideration that consciousness is a subjective experience, and may therefore be nothing more than an epiphenomenon in intelligent beings. This same consideration is echoed in Marcus (2018: p. 127), according to whom, for the development of AI, awareness is not a prerequisite. "It must be an epiphenomenon in humans and possibly other biological creatures. There's a thought experiment that says: could we have something that behaves just like me, but isn't conscious? I think the answer is yes. We don't know for sure because we don't have an independent measure of what consciousness is, so it becomes very difficult to substantiate these arguments."

Koller (2018: 395) goes even further by stating that arriving at General AI or superintelligence (the futuristic predictions of AI) dispenses with consciousness, "because it is possible to have an incredibly intelligent system, which has nothing to do with an inner consciousness." Turing had already stated that consciousness is not knowable. To conclude, faithful to Popper, Koeller considers that, if consciousness is not a falsifiable hypothesis, therefore, it is not science. However, it must be said, this is a good way to escape the problem.

In any case, from these basically convergent considerations, one can reach a conclusion with which I agree, that consciousness is an interior phenomenon, and that, because it is interior, it is difficult to find ways of sharing between consciousnesses, as it is difficult to know whether other animals have it too. On the other hand, being simultaneously interior and exterior, human intelligence is shareable through external signs, being, therefore, collective. Certainly, this flat statement is indebted to many explanations. However, proceeding down this path would take us too far from the question at hand: whether AI is intelligent.

2   A LITANY OF NEGATIVES

Putting machine intelligence into the discussion presupposes reviewing the arguments of those who deny the existence of any type of intelligence in algorithms. Prototypical of this denial is John Searle's position which, moreover, is repeated by all those who continue to follow his thesis. The Chinese Room Argument (Searle 1999), originally published in 1980, has become one of the best-known arguments in cognitive philosophy. Using a kind of thought experiment, the argument concludes that programming a digital computer can make it appear to understand language, but cannot produce real understanding, as computers merely use syntactic rules to manipulate sequences of symbols, but do not have any understanding of meaning or semantics. With this, Searle refuted the thesis, approved by experts at the time, that human minds are computational or information processing systems similar to computers. As for computers, for Searle, they can, at best, simulate human processes, but they understand nothing because they have no consciousness. Today, the argument seems as dated as the idea that the computer is limited to manipulating sequences of symbols.

Kate Crawford, in her sharp criticism of the ills of AI, takes the opportunity of a more popular article (Crawford 2021) to declare, in a sensationalist tone, that "AI is neither artificial nor intelligent. It is made of natural resources, and it is people who perform tasks to make the system appear autonomous." The question remains as to what the author means by artificial and, above all, by intelligent. For other authors, Deep Blue (the supercomputer and software created by IBM especially for playing chess) is just a super calculator and has nothing very intelligent about it: just brute force capable of processing two hundred million positions per second to generate potential solutions for eight future moves in a game of chess. And there are those who, paradoxically, say that AI is exactly what computers do not have (Accoto 2020: 93).

Kremer's (2021) article "Computers don't think, they are oriented in thought" can also be taken as paradigmatic of the deniers of computational intelligence. For the author, the word "intelligence" is used abusively. It is often taken by competence, for example, when a program competently automates a task. "The valorization of competence over intelligence is perhaps tolerable, but the erasure of intelligence in favor of competence is not." Mainly because it would be a shame to miss such a multifaceted word that derives from the Latin "inter" and "legere" whose conjugation can be interpreted in a multitude of ways.

For Kremer (ibid.: 402), "the computer is not a thinking machine, nor can competence be taken for intelligence. The computer does not need to deviate from the realm of possible experience. It is not guided by a subjective need to judge." It is nothing more than a device used to expand the domain of experience by carefully layering mediating representations prepared for experimental use. To function in this space, computers need guidance and require outside assistance—"they don't think for themselves." But the computer "mediates between abstract concepts and reality and therefore serves to guide thought. By guiding thought, machines are much more competent." While human intelligence is a magnetic force for its own compass, computers abandon intelligence in favor of executing commands.

It is impossible not to notice what is behind the author's *parti pris:* intelligence is only human. Any other forms of intelligence are, therefore, corruptions of intelligence or not even intelligence, a problem that is raised by Rus (2018: 263), for which "there is a great misconception in the popular press about what AI is and what AI is not. When they say AI they mean machine learning and, more than that, deep learning within machine learning. People tend to anthropomorphize what these terms mean. Non-experts associate it with intelligence and take it as human intelligence."

3   Agency Without Intelligence?

Bringing the topic to the more recent context of AI, in an article about GPT-3, a third-generation, autoregressive language model that uses deep learning to produce texts similar to human texts, Floridi and Chiaritti (2020: 690) tested the model through mathematical, semantic and ethical questions to conclude, based on the judgment of a publicity articlavene that GPT-3 is "an extraordinary piece of technology, but so intelligent, conscious, smart, attentive, perceptive, insightful, enlightening, sensitive and sensible (etc.) as an old typewriter". If anyone who reads me has ever used a very mechanical, noisy, and tiring typewriter, they will see how abusive the metaphor is, in addition to the immense fog of conceptual mixtures in which the authors involve intelligence, consciousness and other attributes, all very mixed and little explained. But these types of metaphor function in a similar way to clickbait on social media.

In a recent article, written in the heat of the imperious curiosity that ChatGPT 3,5 and 4 are producing, Floridi (2023) advanced his argument for denying intelligence to AI. However, given Chat's undeniable advances in its conversational skills and even tasks, the philosopher granted the Chat the power of agency, a very unprecedented type of agency. Because, of course, for the author, this is an agency devoid of intelligence.

The text is dotted with statements that are becoming clichés, such as those still tied to Searle's theses that AI is merely executive but does not understand the meaning of what it executes. In the face of ChatGPT, along the same lines, it has become commonplace for those who, proud of their humanity, claim that the Chat is nothing more than a parrot of a very special type, a stochastic parrot. All of this deserves to be discussed in more detail. However, I limit myself to asking a question about Floridi's argument: how can there be agency without intelligence?

To discuss this impossibility, I turn to the theory of causality advanced by C. S. Peirce, a philosopher who, unfortunately, is very little followed, due, among other difficulties, to the immense complexity and logical-scientific originality and the transdisciplinarity of his work. If we pay attention to his very original conception of causality, we will find elements capable of refuting the Floridian idea of the possibility of agency without intelligence.

The concept of cause appears in many philosophers over time, which places this concept in the midst of controversy. I won't go into those details. I limit myself to indicating that, for Peirce, there are only two basic actions in the universe: intelligent actions and efficient actions, also called final causation and efficient causation, taking care that they are not confused with the four Aristotelian causes.

So, from his triadic conception of reality, that is, of possibility, action/reaction and continuity as constitutive of each and every phenomenon, Peirce extracted his theory of causality, a theory with three related elements, how could it be otherwise: chance, efficient causation and final causation. Thus, every act of causation involves an efficient component—the concrete action in its here and now –, a final component—the purpose that guides the concrete action—and an element of chance, unpredictable and not determined either by the efficient cause, nor by the final cause. This, in turn, directs concrete processes towards a target, depending on tendencies to achieve purposes. Therefore, all things and people owe their identity to some final cause, which regulates and unifies a series of events, the efficient causes, which constitute momentary states of a continuous process.

Chance, efficient causation and final causation are inseparable. If it were considered in isolation from final causation, efficient, executory causation would be mere blind, brute, unreasonable compulsion. On the other hand, without efficient causation, the purpose, the final causation, would be a pure abstraction, disembodied. It needs the dyadic, efficient relationship between concrete individual events to achieve its goal. Final causation, therefore, is triadic, a relationship between its intended purpose (therefore intelligent), the efficient cause that puts this purpose into action, and the concrete effect that this action achieves. It is important to note that the effect has nothing deterministic, since its real existence, on the one hand, is inseparable from its combination with an infinite swarm of circumstances, on the other hand, it suffers the inevitable effects of chance (Santaella 1999).

Conclusion, there is no agency without intelligence. There it is: without intelligent action as a guide, efficient action would be blind, brutish and ungoverned. Without efficient action, intelligent action would be nothing more than an ineffective abstraction. Both function together, inseparable. They are basic actions of the universe that manifest themselves in each and every phenomenon and certainly in human actions.

Now, what Floridi calls agency without intelligence would be nothing more than efficient action devoid of intelligent action. Transposed to ChatGPT, it would be nothing more than a mindless executor, more ungoverned than a cockroach under the action of Rodox. This is very far from being the case. A case that is none other than recognizing that statistical operations are intelligent operations that AI feeds on. All of this demonstrates that, to understand AI and ChatGPT in their own ways, without imposing typically human parameters on them, both present agencies that do not dispense with purposeful action, which is intelligent action.

The moral of the story is that we need to stop anthropomorphizing AI. One should not continue to block the path of understanding by imposing characteristics on AI that are peculiar to human intelligence, including the absurd demands that AI is not imbued with morality. Now, morality is nothing other than the practical science of ethics, the most imperatively human science. This is the crucial question that ChatGPT is bringing to the table. A type of scene that puts human protagonist in the main focus.

Fortunately, it did not take long for the discourses denying AI intelligence to begin to lose focus, as, increasingly, an affirmative consensus is emerging in which Generativa AI intelligence is taken as a presupposition.

4   A Specific Type of Intelligence

As for the term "artificial", it is used to refer to machine intelligence and not to the broader expression of non-human intelligence that includes animal intelligence for those who are not limited to an anthropocentric view of intelligence. In any case, it is a fact that when it comes to intelligence today it is necessary to be prudent. To begin with, when computer scientists set out to build intelligent systems, they need to know what intelligence means and how it works. It is no wonder, therefore, that the subject of intelligence occupies so much space in their works.

For Russel (2018: 63), an entity can be considered intelligent insofar as it does the right thing, meaning that its actions are expected to achieve its goals. This principle applies to both humans and machines. Doing the right thing is key to the unifying principle of AI. For that, AI needs some key skills, perception, vision, speech recognition, and action. Specifically, it is "the ability to make decisions, plan, and solve problems, as well as the ability to communicate, so understanding natural language becomes very important for AI".

The author adds that "understanding how we know things means entering the scientific field of what we call knowledge representation." This is the way we study how knowledge can be stored and then processed by algorithms that reason, such as automatic deduction and probabilistic inference algorithms. So, we come to learning, a key skill for modern AI. Machine learning means the ability to do the right thing as a result of experience. This requires learning to reason better through experience— "such as figuring out which reasoning steps are useful in solving a problem and which reasoning steps seem to be less useful."

For Betanzos (2023), almost all AI systems exceed our intelligence, but this only happens in a certain field, as most AIs are narrow niche, that is, they are capable of having a very high level of intelligence in a given, very specific field. "They may be great for diagnosing a type of cancer, but they don't work like general practitioners, because the knowledge needed is broader."

In the same vein, Komlosy (2023) states that "it is not advisable to underestimate the strengths of AI applications—namely, that they can process enormous amounts of data much faster and more comprehensively than human minds." The author goes even further: these systems "express forms of intelligence that surpass human intelligence, in the execution of very specific tasks." But it's worth remembering: "comparatively, we shouldn't confuse AI with the creative, non-linear intelligence of human beings, which allows for everything from empathy to irony, cynicism and ambiguity."

Much more recent are some research releases that take away any peace of mind by revealing the IA Generative is far from expressing forms of automated intelligence with a mechanistic content. A. Ramos (2023) explains that AI "is becoming increasingly capable of learning autonomously, without the need for explicit instructions." This is due to "the development of machine learning algorithms, which allow AI systems to learn from raw data, identifying patterns and relationships without the need for human intervention." This phenomenon is called "contextual learning" through which "models learn to perform new tasks based on a few examples, despite not having been specifically trained for them."

Even more worrying are some statements from experts about the prowess of ChatGPT-4, whose AI version is touted as a new type of intelligence that exhibits signs of human reasoning (The New York Times). Sensationalist or not, what comes out of this is the question of whether the industry is building something akin to human intelligence, or else whether some of the industry's brightest minds are letting their imagination get the best of them?

5   The Weights on the Scale

The mention of learning brings us to the key concept of AI, not forgetting that learning is at the heart of the skills that intelligence is made of. This seems to be the Gordian knot that implies that it is wrong to deny intelligence to machines. Yes, AI algorithms are intelligent, and so they are, primarily because they learn, self-organize, correct their mistakes, and reach their targets. It is not by chance that the comparison between human learning processes and those of algorithms has become the most discussed topic among AI specialists, because, certainly, to build intelligent machines, they cannot escape comparison with humans.

It is customary to regard AI as intelligent because the computer has acquired the potential to learn and make decisions based on the data it receives. This does not mean that "this kind of intelligence is no longer decidedly different from ours. Because of its imitation abilities, AI has the quality of identifying informational patterns that optimize trends relevant to work." Also, unlike humans, AI doesn't feel fatigue or sleep. On the other hand, human abilities gain in expansiveness, as, while AI only responds to available data, "humans have the ability to imagine, anticipate, feel and judge changing situations, which allows us to shift from short-term to long term concerns." Human intelligence also does not require a constant flow of external data to function.

Without minimizing all the differences between human and AI intelligence, it is necessary to recognize that, as they are learning machines, they share several ways of learning with humans. In fact, neural networks are capable of learning. To do this, they make adaptive changes to the weights and, sometimes, also to the connections, and learning modifies the weights. They learn in different ways: supervised, unsupervised, reinforcement and semi-supervised learning. They therefore share ways of learning that are also human.

In 1977, in her anthological book on *Artificial intelligence and natural man*, Boden sought comparisons between the ways in which humans and machines learn. For her (p. 247–249), there are three ways of generating new thoughts: learning, creativity and problem solving. The three are not separated into drawers. Thus, general learning may derive from specific experiences and may involve creative thinking. Likewise, the spontaneous construction of new representations may be instigated by a particular need or problem, and may be aided by environmental cues. Solving a specific problem may require creativity and lead to general knowledge.

Next, the author explains each of the three types of learning she studied: learning by example, learning by hearing and learning by doing. The three respectively bring new knowledge of clues and models, new knowledge of facts, and new skills. Sometimes, learning occurs when new information is accepted, integrating it into a previous structure. Other times it is about reorganizing information that is already in the mind. It is difficult to say whether one type of learning is superior to the other. In any case, learning through examples is the way of learning that both humans and machines perform. Everything depends on a gradual improvement of representations of the world that is instigated by the training of better-structured sequences in particular ways (Boden 1977: 264). Therefore, learning through examples does not mean directly apprehending reality, distorted by an intermediate interpretative activity. In the case of humans, it involves discernment in developing descriptions, or interpretative schemes representing the target domain, schemes which, moreover, are constantly checked by reference to examples or counterexamples so that salient clues can be identified.

As for learning just by hearing, both in machines and humans, learning must be able to tolerate vague epistemological criteria as much as have a richly equipped memory and powerful inferential (i.e., problem-solving) competence in terms of reading between the lines. This implies doing something, that is, thinking about an activity that machines perform to a certain extent and with due proportions. Learning by doing is most commonly understood as learning to do something by trying to do it, or learning to do it better through the repetitive practice of which machines are masters (ibid.: 278).

Forty-six years have passed since this publication by Boden, and a lot of water has flowed in the field of AI until we have reached the specific success, in recent years, of machine learning and, in particular, of deep learning. Thus, supervised learning is very similar to human learning that occurs under the supervision of a teacher. The teacher presents some specific examples for the student to memorize, and the student then derives general rules from them. But human and animal learning is largely unsupervised: we discover the structure of the world by observing it, not by being told the name of each object. This learning is found in the methods used by humans to determine that certain objects or events belong to the same class and how this happens when observing the degree of similarity between them (Mueller and Massaron 2020: 128).

Reinforcement learning is well known as it constitutes a chain of psychological studies, behaviorism, used both in research and in therapeutic and educational practice. The explanation given to us by Sejnowski (2019: 172) about this way of learning is so suggestive that it deserves to be mentioned here. The author gives the example of reinforcement learning in the parallel between the way birds learn to sing and children learn to speak. In both cases, there is an initial period of auditory learning that anticipates a later period of progressive motor learning. Sejnowski demonstrates his thesis with zebra finch birds that hear their father's song as soon as they are born, but do not produce sounds of their own until months later. "Even when they are isolated from their father, before the motor learning phase, they go through a period of experimentation with sounds that continues to improve and ends up crystallizing the bird song in their father's dialect." Zebra finches know which part of the forest a member belongs to just by listening to their song, just as we know where a person is from by their accent. The hypothesis that has driven research into birdsong is that, during the auditory learning phase, a model is learned and then used to refine the sounds produced by the motor system in the subsequent phase. In both humans and songbirds, "the pathways responsible for the learning phase are in the basal ganglia, where we know reinforcement learning occurs" (ibid.).

6   IMAGINATION, CREATIVITY AND EMOTIONS

Comparative weights in the scales of human intelligence and AI end up, as a rule, leading to the themes of imagination, creativity and emotions, which function as frontiers to defend the limits that differentiate human intelligence from AI and which, as it is argued, AI will never be able to overcome. Imagination is a territory of study that has accompanied philosophy, anthropology, psychology, cognitive semiotics and literary studies. Although it has a connection with the word image, the terms imagination or imaginary are always more linked to the domain of mental images. This is how the concepts of imagination or imaginary appear in the theories of the most notable authors on the subject, such as Sartre ([1940] 1978), who conceived it as an intentional act of consciousness. In the celebrated work *Anthropological Structures of the Imaginary*, Gilbert Durand (2002) also privileges the imaginary as one of the human being's mental faculties. For him, all thought rests on general images, archetypes, which function as unconscious determinations of thought.

Another author who was notable for building a theory of the social imaginary, in the field of sociology, was Cornelius Castoriadis (1975). For him, the imaginary is the key to thinking about collective phenomena. Societies are characterized as a set of imaginary social meanings that are embodied in institutions in which these meanings gain external life. These examples are enough to perceive a marked tendency to consider the imaginary as a faculty that comes from within the human mind. This constitutes a convergent point, despite the theoretical differences between the authors.

Considered in this way, machines are, in fact, very far from being able to compete in terms of the imagination that humans are so proud of. It is no surprise that Pearl (2018: 367), a specialist in AI, places imagination among the obvious limitations of AI, in which it is not possible to think about something that has not been seen before, in the sense of seeing, intervening and to imagine. "Imagining is the top level that requires counterfactual reasoning: what would the work be like if it were done another way? They are imaginary scenarios, a game of creative pretending." Faced with this limit, the author defends the need to build new models of the world. "Imagining a world that does not exist gives us the ability to come up with new theories and also to repair our old actions in order to assume responsibility, repentance and free will. Worlds that don't exist, but that could exist." Transposed to AI, understanding this logic means being able to build "machines that can imagine things, that take responsibility, that understand ethics and compassion. I'm not a futurist and I try not to talk about things I don't understand. But I understand how important counterfacts are in every cognitive task that could be implemented on a computer." To complete, the author declares that she has some drafts of how to program free will, ethics, morality and responsibility into a machine. This would, without a doubt, be an ideal path. So, let's wait.

Creativity is the second barrier between humans and machines. The literature on human creativity is immeasurable and therefore, at this moment, for us, unreachable. Fortunately, what interests us is the relationship between us humans and machines. In 1977, Boden dedicated almost a hundred pages of her book to the theme of creativity in the context of AI, limited to the state of the art in that period. The author returned to the topic in 2020 (pp. 77–78), discussing three types of human creativity: combinatorial (known ideas combined in an unusual way, producing a statistical surprise), exploratory (exploitation of a culturally valued way of thinking) and transformational (generation of new structures).

With the intention of highlighting human creativity in the face of machines, for Aoun (2017: 48–50) to think creatively is to think divergently, divergence being defined as the creative generation of multiple responses in a free flow of ideas. There are, however, some common points between convergent and divergent thoughts, namely, both require the ability to access and elaborate. But divergences require creativity, which, in turn, is defined as "a sensitivity to the changing nuances of a problem, a facility for reformatting it according to the demands of circumstances and, finally, the ability to generate a result or resolution that contains elements that weren't there at the beginning."

After the emergence of great language models and the Transformer neural network, AI models became capable of writing texts. However, divergent thinking does not appear in them, as the human brain is still necessary for this. Despite this, maintaining the view that the mastery of facts and knowledge is what makes a person intelligent or prepared is nothing more than a disproportionate view of human intelligence, especially at the present time, when robots, advanced machines and AI in general they are increasingly mastering facts and knowledge as effectively as humans.

Even when considering critical thinking in the competent analysis of ideas that humans are capable of and applying them in a fruitful way, it must be considered that machines are certainly improving their capabilities also in this respect. Observing, analyzing and communicating are powers that are advancing in machines, but, for now, they still do not have the ability to synthesize and imagine. While machines are better than humans "at using data inputs to solve a specific problem—winning a game of chess, organizing a global supply chain, finding you a date for a Saturday night—they are not as impressive when it comes to non-quantifiable thinking" (Aoun, ibid.: 62).

Machines are adept at understanding elements in complex systems and the ways their variables intertwine, but they are less skilled at understanding how to apply this information in different contexts. To demonstrate this, Aoun discusses the example of a machine with the potential to "model the impact of climate change on a coastal area, accessing water temperature, pollution, currents, water patterns, and a set of intertwined factors." By accessing all this data, this machine can come to conclusions about how to improve the surrounding architecture and combat erosion. But that same machine wouldn't be able to imagine how to deploy the data in different fields like economics, law or health sciences. Conclusion, "computers can be programmed to think across a variety of silos, which enables them to engage in systems thinking of a certain kind, but the great creative leaps that occur in humans are still unattainable for machines" (Aoun 2017: 65).

Finally, a territory full of ambiguities arising from its complexity is that of emotions. Are machines sentient? Are they able to feel? When talking and non-talking robots guess our thoughts and respond to them with warmth and apparent care, aren't they feeling it? Our facial expressions are a window into the emotional state of our brain. When deep learning can see through that window, isn't the machine feeling it?

According to Sejnowski (2019: 195), traditionally, cognition and emotion were treated as separate functions of the brain. In general, it was thought that cognition was a cortical function, and emotions were subcortical. "In fact, there are subcortical structures that regulate emotional states such as the amygdala, which performs a biological function when emotional levels are high, especially fear; but these structures interact strongly with the cerebral cortex." Thus, "the engagement of the amygdala in social interaction, for example, will lead to a stronger memory of the event." Not only does it follow that "cognition and emotions are interconnected," but it also becomes possible to evaluate the role that emotion plays in the functioning of the mind as a whole, especially in memorization.

Pearl (2018: 371) informs us that we have chemical fluctuations in our body, and they have a purpose. "The chemical machine interferes and, at times, surpasses the reasoning machine in the face of emergencies. Thus, emotions are just a chemical machine for adjusting priorities." Furthermore, our intelligence is coordinated with the central nervous system, the visual motor system and modalities of cognition that involve planning, reasoning, emotions, intention and persistence. None of this is seen anywhere in AI systems (Li 2018: 149).

A few years ago, computer scientists and even philosophers and psychologists, when discussing intelligence, did not think of it as something that requires emotion. However, more recently, researchers have begun to emerge who seek to think about the mind as a whole, which cannot fail to include emotion (Boden 2020: 107). The ambition of scientists is to model and simulate states of mind and emotions. Even if they do, the algorithms still lack hormones and neuromodulators. More than that, even if the ambition of modeling hormones and neuromodulators is achieved, the algorithm will still lack acting on a living body, situated in a dynamic environment and actively engaged in it. The environment and engagement are both physical and sociocultural. In this context, the fundamental psychological properties are not reasoning or thinking, but adaptation and communication (ibid.: 187).

7 Ending Balance

The considerations that were selected to appear in this final balance have already been, to a certain extent, addressed in Santaella (2023: 256–265). With some modifications, they return here for their synthetic power. It is not only customary to deny, but also to affirm the existence of some level of intelligence in AI. This chapter adopted the second hypothesis and worked to defend it. In fact, AI is intelligent because the computer has acquired the potential to learn and make decisions based on the information it receives. However, we are led to agree with Cremer and Kasparov (2021) that this does not mean that "this type of intelligence is no longer decidedly different from ours. This is an extremely useful type of intelligence in an organizational environment," given that "AI has the quality of identifying informational patterns that optimize trends relevant to work." Furthermore, unlike humans, AI does not experience fatigue, sleepiness or procrastination. On the other hand, however, human abilities gain in expansiveness, because, while AI only responds to available data, "humans have the ability to imagine, anticipate, feel and judge changing situations, which allows us to shift from short term to long term." Human intelligence also does not require a constant flow of external data to function.

Unfortunately, when referring to machine learning, people imagine that the machine learns just like humans do. But these terms mean very different things in a technical context. "These systems do better than humans because they can assimilate and correlate many more data points than humans can." But when the system learns that there is a cat, for example, in a photograph, "what it is really doing is saying that the pixels that make up this figure, which humans have labeled as a cat in the photo, are the same as other figures that humans have labeled as a cat in the photo." But the system doesn't have the vaguest idea of what a cat represents (Rus 2018: 263).

As discussed above, in her book on *Artificial intelligence and natural man*, Boden (1977: 247–389) presents several forms of learning. If we consider the stage of development that AI is in today, some forms are common between AI and humans, but others act as bottlenecks for AI. For example, learning from examples in humans involves discernment and development of descriptions, or interpretative schemes, whereas in machine learning the ability to learn through a very large number of examples is its greatest asset. The human ability to learn by doing, however, which is at the basis of the practice that leads to the repetition and ever-increasing improvement of that practice, presents principles that the black box of deep learning must also carry out, although this cannot be said with certainty, given the fact that there is no clear understanding of the logical movements that are processed in the layers of neural networks. But, among the major bottlenecks of AI, which are pointed out by experts, is the human capacity for abstraction, which is the basis, among others, of diagrammatic reasoning, driven by the logical imagination of iconic thinking.

The weights on the scales of human and AI intelligence, which were compared in this chapter, reveal that they are two distinct types of intelligence. For Etzioni (2018: 497), humanity is focused on "building a model that can be inspected, debated, explained and improved. Without being able to explain its reasoning, in turn, the machinery tells us: trust us because we are more often right than wrong." There are still many limitations of AI and all confrontations between AI and humans are essential to put an end, at least for now, to the mistaken fears that AI is devouring human intelligence, precisely because, when human and Artificial Intelligence are placed in confrontation, a paradox arises: what is difficult for humans, AI does; what is difficult for AI, humans do. In reality, as they are two distinct types of intelligence, however much human intelligence may or may not make us proud, this does not allow us to deny the cognitive potential and expression of intelligence in AI. Evidently, there are characteristics and properties that act as obstacles for both to be equal.

Machine systems do not have their own internal purposes (Tenenbaum 2018: 484). This is the current state of the art of research that is in development, but this does not justify people touting litany of hyper-valuation of human intelligence to the detriment of machines, because, in science, what is not possible today, could be possible at some point, as long as the ethical constraints are taken into account. It seems sobering to conclude, with Lecun (2018: 133), that AI systems will amplify human intelligence in the same way that mechanical machines were an amplification of physical strength. They will not be a replacement. "Just because an AI is capable of recognizing the image of a tumor does not mean that radiologists will be relieved of their work. It will be a very different job, and much more interesting. They will do something more interesting like talking to patients instead of watching screens eight hours a day."

In short, "we must foresee a future in which computational systems will have abilities complementary to those of humans" (ibid.: 340), because, according to Cremer and Kasparov (2021), "when put to work together, both intelligences are complementary, passing to be called 'augmented intelligence', a type of collaborative intelligence, when it involves a collaborative effort in the service of humans," that is, a type of collaboration in which technology and humans go hand in hand (Mueller and Massaron 2020: 207). Domingos (2017: 69) joins the chorus when he states that "the future belongs to those who know, at a very deep level, how to combine their area of specialization with what algorithms do best." Unfortunately, votes of hope alone have not been able to erase the negative externalities or side effects of AI, which places a responsibility on human shoulders that cannot be escaped.

It can be concluded that AI systems can amplify and will continue to amplify human intelligence in the same way that mechanical machines were an amplification of physical strength. They will not be a replacement. In short, we must foresee a future in which computer systems will have abilities complementary to human skills, functioning much more like a new type of augmented intelligence in the interchange and union of two distinct types of intelligence.

In any case, penetrating the core of AI means ratifying the fact that the human is a unique creature. We have a set of talents that makes us unique among animals and now also unique in confronting the advances in AI. The plasticity of human mind and behavior makes this uniqueness possible. This is what characterizes us. "Humans are unique not only because they do science, nor because they do art, but because science and art are equally expressions of the wonderful plasticity of our mind. The brain and a baby precisely represent the point where the plasticity of human behavior begins" (Bronowski 1973: 412). It can be concluded that AI systems can amplify and will continue to amplify human intelligence. They will not be a replacement. In short, we must foresee a future in which computer systems will have abilities complementary to human skills, functioning much more like a new type of augmented intelligence in the interchange and union of two distinct types of intelligence.

REFERENCES

Accoto, Cosimo. 2020. *O mundo dado. Cinco breves lições de filosofia digital*, Eliete da Silva Pereira (trad.). São Paulo: Paulus.

Aoun, Joseph E. 2017. *Robot-proof. Higher education in the age of artificial intelligence*. Cambridge, Mass: MIT Press, 2017.

Betanzos, Amparo Alonso. 2023. Inteligência artificial pode superar a humana? 8 perguntas sobre a tecnologia. A IA pode ter consciência própria e sentimentos? Entrevista concedida a Alicia Hernández, 2023. Available at: https://www.bbc.com/portuguese/articles/czvp8ypwqz9o?fbclid=IwAR1KTSE4Wv9TheVoaghd6YZYiZvaj6Uzjzh89TO5L8Sg9wwEkHbZeR0zzY. Access: 08/06/2023.

Boden, Margareth A. 1977. *Artificial intelligence and natural man*. Brighton: The Harvest Press.

Boden, Margareth A. 2020. *Inteligência artificial. Uma brevíssima introdução*, Fernando Santos (trad.). São Paulo: UNESP, 2020.

Bostrom, N. 2018. Interview to Martin Ford. In: Martin Ford (ed.). *Architects of intelligence. The truth about AI from the people building it*. Birmingham: Packt Publishing, pp. 97–116.

Bronowski, J. 1973. *The ascent of man*. London: Science Horizons Inc.

Castoriadis, Cornelius. 1975. *La institución imaginaria de la sociedad*. Barcelona: Tusquets Editores.

Chalmers, D. 1995. Facing up to the Problem of Consciousness, *J. of Consciousness Studies* 2: 200–219.

Crawford, Kate. 2021. AI is neither artificial nor intelligent. The Guardian, 2021b. In: https://www.theguardian.com/technology/2021/jun/06/microsofts-kate-crawford-ai-is-neither-artificial-nor-intelligent#:~:text=AI%20is%20neither%20artificial%20nor%20intelligent.,make%20the%20systems%20appear%20autonomous.&text=Problems%20of%20bias%20have%20been%20well%20documented%20in%20AI%20technology. Acesso: 10/08/2021.

Cremer, David de, Kasparov, Garry. 2021. AI Should Augment Human Intelligence, Not Replace It. Harvard Business Review, In: https://hbr.org/2021/03/ai-should-augment-human-intelligence-not-replace-it. Access: 24/01/2022.

Domingos. 2017. *O algoritmo mestre*, Aldir José Coelho Corrêa da Silva (trad.). São Paulo: Novatec.

Durand, Gilbert. 2002. *Estruturas imaginárias do imaginário*. São Paulo: Martins Fontes.

Etzioni, Oren. 2018. Interview to Martin Ford. In: Martin Ford (ed.). *Architects of intelligence. The truth about AI from the people building it*. Birmingham: Packt Publishing, p. 493–509.

Floridi, Luciano e Chiriatti, Massimo. 2020. GPT-3: Its Nature, Scope, Limits, and Consequences. *Minds and Machines* 30 (4), p. 681–694.

Floridi, Luciano. 2023. AI as agency without intelligence: on ChatGPT, Large Language Models, and other Generative Models. *Philosophy & Technology*, v. 36, n. 15, p. 1–15.

Koller, Daphne. 2018. Interview to Martin Ford. In: Martin Ford (ed.). *Architects of intelligence. The truth about AI from the people building it*. Birmingham: Packt Publishing, p. 387–403.

Komlosy, Andrea. 2023. What lies beyond the AI tipping point, 2023. Entrevista concedida ao *Le Point*, 09/02/2023. In: https://www.project-syndicate.org/onpoint/ps-commentators-respond-what-lies-beyond-the-ai-tipping-point. Access: 09/02/2023.

Kremer, Attay. 2021. Computers do not think, they are oriented in thought. *AI & Society*, 36, p. 401–402.

Lecun, Yann. 2018. Interview to Martin Ford. *In*: Martin Ford (ed.). *Architects of intelligence. The truth about AI from the people building it*. Birmingham: Packt Publishing, p. 119–142.

Li, Fei-Fei. 2018. Interview to Martin Ford. In: Martin Ford (ed.). *Architects of intelligence. The truth about AI from the people building it*. Birmingham: Packt Publishing, p. 145–160.

Marcus, Gary. 2018. Interview to Martin Ford. In: Martin Ford (ed.). *Architects of intelligence. The truth about AI from the people building it*. Birmingham: Packt Publishing, p. 305–330.

Mueller, John Paul, Massaron, Luca. 2020. *Inteligência artificial para leigos*, Alberto Gassul Streicher (trad.). Rio de Janeiro: Alta Books.

Pearl, Judea. 2018. Interview to Martin Ford. In: Martin Ford (ed.). *Architects of intelligence. The truth about AI from the people building it.* Birmingham: Packt Publishing, p. 357–373.

Ramos, Ademilson. 2023. A Inteligência Artificial já está aprendendo de forma autônoma, 2023. Engenhariae. Disponível em: https://engenhariae.com.br/tecnologia/a-inteligencia-artificial-ja-esta-aprendendo-de-forma-autonoma-dispensando-a-necessidade-de-instrucoes-explicitas?fbclid=IwAR3kPKEifi-Sn4-Lo6OJ6OO0epsNySycTJGtCE1zJOTMKJtoJj13-iM55ss. Acesso: 20 jun, 2023.

Rus, Daniela. 2018. Entrevista concedida a Martin Ford. In: Martin Ford (ed.). *Architects of intelligence. The truth about AI from the people building it.* Birmingham: Packt Publishing, p. 253–268.

Russel, Stuart J. 2018. Interview to. In: Martin Ford (ed.). *Architects of intelligence. The truth about AI from the people building it.* Birmingham: Packt Publishing, p. 39–68.

Santaella, Lucia. 1999. A new causation for the understanding of the living. *Semiotica* 127 (3–4), p. 481–496.

Santaella, Lucia. 2023. *A inteligência artificial é inteligente?* São Paulo: Almedina, 2023.

Sartre, J. P. *L'imaginaire*. 1978. Paris: Gallimard, 1940. Tradução para o português: *O imaginário*. São Paulo: Abril Cultural.

Searle, J. 1999. The Chinese Room. In R.A. Wilson and F. Keil (Eds.), *The MIT Encyclopedia of the Cognitive Sciences*, Cambridge, MA: MIT Press.

Sejnowski, Terrence J. 2019. *A revolução do aprendizado profundo*, Carolina Gaio (trad.). Rio de Janeiro: Alta Books.

Tenenbaum, Joshua. 2018. Interview to Martin Ford. In: Martin Ford (ed.). *Architects of intelligence. The truth about AI from the people building it.* Birmingham: Packt Publishing, p. 473–491.

CHAPTER 3

Anthropomorphizing and Trusting Social Robots

Pietro Perconti and Alessio Plebe

1 SMALL INTRODUCTION

The chapter explores the challenges posed by the proliferation of robots, particularly humanoid robots, in society, focusing on cognitive constraints, ergonomic concerns, cross-cultural issues and the emergence of an artificial morality in human–robot interactions. It also delves into the distinctions between physical and intentional trust, highlighting the role of deference, both biologically grounded and epistemic, in human–robot relationships.

The concept of selective deference is introduced as a means of navigating interactions with robots, emphasizing the importance of balancing trust and deference based on the context and the technology involved. Overall, it is underscored the need for a nuanced understanding of trust and deference in the evolving landscape of human–robot interactions.

P. Perconti (✉) • A. Plebe
University of Messina, Messina, Italy

P. Alexandre e Castro (ed.), *Challenges of the Technological Mind*, New Directions in Philosophy and Cognitive Science, https://doi.org/10.1007/978-3-031-55333-2_3

2   BECOMING FAMILIAR WITH ROBOTS

Over the past decades, our everyday life has been crowded with an ever-increasing number of robots. Many kinds of robots are more and more available, including the vast field of industrial automation. However, it is humanoid robots that are having the most significant influence on people's ordinary lives. There are humanoid robots that serve as receptionists, interactive info points, waitresses in restaurants, nurses, and companions for a romantic or sexual relationship.

The landscape in the near future could surpass our imaginations, driven by the synergy between the fields of humanoid robotics and large neural language models. Ongoing research is focused on equipping robots with the ability to engage in linguistic interactions with humans and to reason about their tasks, leveraging Transformer-based neural language models.

It is now no longer just a matter of facing the idea of an Artificial Intelligence (AI), but rather one of dealing with intelligent bodies which are engaged in routines of social cognition with humans. All this presents challenges that we had not addressed until now. There is a widespread feeling that we are dealing with an urgent situation, that is, a new social circumstance that requires both innovative analysis and consequent public measures. There is a social concern about the rules to be imposed on the field of Artificial Intelligence and the resulting ethical challenges it poses. Reactions range from a stand inspired by a sort of "no special regulation" for AI to those who envision special legislation for information technology and even proponents for temporary or permanent bans.

The issue that will be addressed here concerns the cognitive constraints that influence the relationship between people and humanoid robots. We will try to understand what makes this relationship more or less ecological and how such ecological constraints should be taken into account in designing these kinds of devices. The question we would like to address here is: how can people become familiar with such artifacts? And, what kind of problems does such familiarity raise? We would like to argue in favor of the idea that the appeal to constructive anthropomorphism (Bruni et al. 2018) could be useful in addressing the problems just mentioned. Although anthropomorphism is generally seen as a risk for scientific investigation about the relationship between people and other animals and a bias when directed toward inanimate objects and technological devices, in reality it can—if used consciously and prudently—be a significant resource (Epley et al. 2007).

It is a matter of investigating the mechanisms by which a trusting relationship between humans and objects can be established. In human evolutionary history, this is perhaps the first time that we have dealt with trusting an object in a way that is not metaphorical. Of course, even in the past, citizens of a village, for example, had to decide whether to trust the time measure of the church steeple. Up to the present day, however, basing the achievement of their own purposes on relying on the workings of a machine or the internal states of another person have been two radically different things. The difference is grounded on the idea that the minds are semantically opaque creatures, that is, they are depending on intensional logic, while machines are semantically transparent. When, however, we come across humanoid robots the usual compass no longer seems to work. In a sense they are undoubtedly inert matter, like usual objects. First of all, they are not living organisms; they do not have any interest in the world that is regulated by hunger, thirst, survival and reproduction instincts. Humans spontaneously care if the air in the room they are in is too stale. Robots don't care. Or, at least, they don't care unless they are equipped with appropriate sensors and instructions to exhibit a behavior that is susceptible to be interpreted as a concern for stale air.

We can retain the distinction between minds, understood as semantically opaque creatures, and machines, conceived as semantically transparent objects. Or we can revise that distinction because of the encounter with humanoid robots. As long as it had been a matter of considering old-fashioned mechanical automata, human appearance had been merely a source of curiosity and amusement. But since those automata accommodate within them an assortment of sensors and a computer equipped with a deep neural network, things have changed. Those objects that had always seemed transparent to us have become opaque, exactly like minds.

3   ERGONOMIC, CROSS-CULTURAL AND MORAL ISSUES

Over the last years, the interaction with humanoid robots is raising many kinds of trouble. The major types are ergonomic, cross-cultural, and moral. Ergonomic problems, in turn, are of two different sorts, depending on whether we are considering classical ergonomics or cognitive ergonomics. Classical ergonomics is about trying to adapt objects to the way the human body is made or, at least, to take into account certain bodily constraints that influence the interaction between the way the human body is made and the objects it happens to run into. The need for cognitive ergonomics arises alongside classical ergonomics, which, by contrast, can also be called "physical" ergonomics. While classical ergonomics deals with the adaptation of objects to the human body, cognitive ergonomics focus on the psychological aspects of interaction with artifacts. Jacques Carelman's concept of "impossible objects", or *objets introuvables*, shows how we often take it for granted that there must be a minimal agreement between everyday objects and the human body, similar to George Moore's truisms of common sense, which we are forced to take for granted, at least as long as we wish to share the same form of life with other people (Carelman 1969; Moore 1925).

In contrast, cognitive ergonomics deals with the psychological aspects of interacting with artifacts. Cognitive ergonomics considers objects as having their own psychology, as if they have emotions and the mental states which are typical of folk psychology, such as beliefs and desires. Apparently for humans, objects, including artifacts, possess their own "mind," i.e., they regularly appear to us as entities endowed with psychological features. In this way we find certain cars to be more "bad" or "aggressive" than others, because of their appearance even before their performance. The interface of certain technological devices sometimes seems "hostile" to us and other times "friendly." It can even happen that we find ourselves swearing at a riotous printer. Cognitive ergonomics has any ontological commitment to the minds of cars or printers. It is simply based on a psychological tendency that is enough when shared to suggest that this should be taken into account in designing those artifacts that are intended to perform their functions in constant interaction with people.

A second problem comes from cultural differences around the world. It is very common to note how varied are the ways in which people react to electronic devices, such as mobile phones or cameras. In certain countries across the world, for example, some people do not like to have their picture taken. Something similar happens in the case of humanoid robotics. There is an interesting difference in the reactions that humanoid robots trigger in Western and Eastern cultures. While the science of robotics is equally developed worldwide, humanoid robotics is characterized by a strong Japanese supremacy, and Eastern robotics in general. In Japan, relations with humanoid robots are generally well respected. Projects aiming at building humanoid robots capable of caring for other people, especially the elderly and sick, are strongly encouraged. The reason for this different social consideration of humanoid robots in the East and the West can be questioned. In the absence of extensive research on the subject, one could perhaps only make some speculations inspired by a variety of cultural stereotypes. It can be assumed that the well-known preference given by the Japanese people to impersonal relationships, with the typical massive use of vending machines and any electronic devices bypassing "emotionally hot" human relationships, plus the typical Japanese discretion toward the most intimate and personal sphere, results in elderly and sick people preferring to deal with machines in the more embarrassing aspects of care relationships. But, of course, this is just speculation, perhaps overly conditioned by the most common cultural stereotypes.

The third type of issues is ethical and results in a relatively new field of scientific investigation, that is, artificial morality. It is the very idea of an "artificial morality" that at first glance could appear rather weird. Because it consists of judgments and intentions, morality seems, at first glance, to be a specifically human creature.

Morality exists only where there is a human environment, which consists of language, intentional vocabulary and judgments about the world. Overall, two problems can be distinguished in the field of artificial ethics. The first concerns the question: What are the ethics needed to govern relations with machines that have intelligence?—For example, do we need an ethics which is also toward things, besides persons? The second type of problem concerns the question: What ethics should machine with intelligence and self-awareness be equipped with?

The first question lacks any ontological commitment to machine morality. It simply concerns the kind of ethics we should adopt to govern our dealings with machines that are assumed to be intelligent, or whose behavior appears to be endowed with some form of intelligence. The story goes as follows. If the machines we are dealing with show no signs of intelligence, the question of their ethical evaluation does not arise at all. They are simple tools, like hammers or classic household appliances. A blender, an electric stove or a refrigerator does not require moral consideration. They are simple tools that we use when we are interested in their functions. But, if at some point, it seems that an appliance needs to make a decision, then that will change. What if blenders were designed to refuse to grind meat?

Suppose the designers were vegans and radical animal rights activists, and we found a way to prohibit their blenders from doing to meat what they normally do to vegetables or inert substances. Would we still look at blenders in an ethically neutral way, or wouldn't some of the ethical considerations we apply to designers who advocate for animals also apply to their blenders? So perhaps a vegan activist might feel comfortable with such a blender, while a burger lover would not be pleased to have it in their home. If this kind of reasoning has its own plausibility with a simple blender that may not grind meat, consider how the need for ethics capable of regulating relationships with intelligent machines will become increasingly urgent as AI is distributed in machines.

In addition to these two types of problems, there are also two different approaches used to address the above issues: top-down and bottom-up approaches: "Top-down approaches involve implementing explicit theories about moral behavior in algorithms. Bottom-up approaches involve attempts to train or develop agents whose behavior emulates morally praiseworthy human behavior" (Allen et al. 2005, p. 149). Top-down approaches start from a pre-existing ethical framework. So let us assume that someone is either a utilitarian or a Kantian committed to a deontological ethics. Based on his or her own biases, the utilitarian or Kantian will try to specify a set of obligations that intelligent machines should obey; in other words, he or she will try to specify constraints that the designers of intelligent machines, who are also moral agents, must obey if they are to design artifacts that are socially acceptable. The goal can be achieved either by building utilitarian or Kantian machines, or by designing machines whose behavior is compatible with utilitarian or Kantian ethics. Bottom-up approaches, on the other hand, emphasize the development of moral sensibility through machines, and take a stance similar to evolutionary theory, imagining the path that machines should take to acquire the moral sensibility that humans have gained through millennia of biological evolution and nurture. Such an approach can be found in developmental robotics, which has moved away from the challenge of developing cognitive skills, but is rapidly expanding into the development of moral skills (Cangelosi and Schlesinger 2015).

4   Trust and Deference in Robots

Within the context of the challenges mentioned earlier, the issue of trust takes center stage as a complex and multifaceted problem is central. Authenticating trust in a robot, even one that is meant to mimic human appearance and behavior, presents a number of challenges. It often feels unnatural because it means trusting a machine, a being without consciousness or emotion—a stark contrast to the trust we naturally place in our fellow humans. This dichotomy in our approach to trust toward robots and humans underscores the complicated nature of our evolving relationship with advanced technology. In Western societies, the prevailing attitudes toward robots are marked by a certain ambivalence. On the one hand, there is a growing desire to harness the capabilities of robots and use them as tools to streamline tasks, increase productivity, and even perform dangerous or mundane tasks that humans would prefer to avoid. This attitude tends to treat robots as subservient beings designed to serve our needs and desires.

On the other hand, there is an undercurrent of fear and skepticism that is fueled by the portrayal of robots in popular culture as a potential threat to humanity. This fear stems from the notion that robots might one day overstep their programmed boundaries, leading to unintended consequences and perhaps even turning against their creators. It should be noted that these two extreme attitudes toward robots—the desire for domination and the fear of rebellion—are not conducive to harmonious and productive coexistence between humans and robots. Instead, these attitudes can perpetuate a sense of unease and distrust in society. Given these challenges, a more constructive approach seems to be to build a relationship based on cooperation. Cooperation should be the cornerstone of our interaction with robots (Plebe and Perconti 2022). However, for this vision to become a reality, we need to look more deeply at what it really means to trust or give responsibility to robots. Collaboration requires mutual understanding and a certain level of trust. In order to collaborate with a robot, people must have confidence in the robot's ability to reliably and safely perform the tasks assigned to it. This trust should be rooted in a clear understanding of the robot's capabilities and limitations, as well as the scope of its tasks. It requires that we define the boundaries of trust and ensure that we do not over-rely on robots in situations where human judgment and oversight are essential.

Moreover, trust in robots should go beyond mere functionality. It should also include ethical considerations and ensure that robots follow a set of ethical principles that are consistent with societal values. Ethical programming and the incorporation of empathy and compassion into robot behavior are essential components of this trust-building process.

Navigating the complex terrain of trust in robots is crucial as we continue to integrate advanced technology into our lives. Striking a balance between our desires, fears, and the imperative for cooperation is essential for a harmonious coexistence with robots. Achieving this balance requires a nuanced understanding of what it means to trust robots and the development of ethical guidelines to guide our interactions with these increasingly prevalent technological companions.

The kind of trust we place in our fellow humans depends on our reliance on their intentional states. This is different from relying on the operation of an elevator, where what really matters is the design and the laws of physics. Humanoid robots are creatures on the borderline. Insofar as they are "robots", in the sense of mechanical automata, they elicit the kind of reliance we place on elevators. But as they are "humanoid", that is, sprinkled with natural affordances that spontaneously lead our brains to activate social cognition, they evoke the kind of trust we generally reserve for rational agents. "Intentional trust" is subjected to different rules than "physical trust." Essentially, while the former is governed by folk psychology, the latter is based on folk physics (which, at least in the case of elevators, is hopefully consistent with experimental physics).

What we are suggesting is that robots deserve intentional trust. They do, in fact, exhibit a number of both physical and psychological features enough to elicit the intentional stance (*à la* Dennett). They are endowed with the typical physical features that we have elsewhere called "mental triggers" (Plebe and Perconti 2022), that is, those perceptual patterns which trigger social cognition in the brain (Graziano 2013). Typically, humanoid robots have faces, try to simulate a type of biological motion and try to be reactive to human gaze. These are the cues that the environment offers to the human brain and which the latter interprets as signals that it is probably dealing with a rational agent. As the appearance of humanoid robots is improved in terms of ecological patterns appropriate for eliciting the neurobiological mechanisms that govern social cognition, to the same extent humans will find it spontaneous to regard such creatures as rational agents.

In such a natural tendency of the human brain lies an extraordinary opportunity for the technological development of humanoid robotics. It is a matter of making the interaction between humanoid robots and the human mind as ecological as possible by building on the achievements of cognitive psychology and other areas of cognitive science. Constructive anthropomorphism can play a useful guiding role in such an attempt. It involves specifying the environmental parameters that elicit social cognition more fully and precisely than the initial suggestion of mental triggers allows. Once these parameters are then computationally modeled, they can be implemented in the body of the humanoid robot so as to act as a set of ecological cues for the human brain.

In this way there will be a better agreement between the body of the humanoid robot and the biology of the human brain. The tendency of humans to attribute human-like characteristics to non-human beings can be a productive strategy for fostering lasting connections between humans and human-like automated virtual agents (Nof 2009). In certain situations, interacting with a human-like virtual agent may prove more effective than doing it with a human. For example, virtual therapists could elicit more openness from patients in some cases compared to human therapists.

It has also been experimentally tested (Christoforakos et al. 2021) how constructive anthropomorphism can play a guiding role in the design of humanoid robots. It was investigated how anthropomorphism affects other possible determinants in the interaction between robots and humans. The other factors considered were competence and warmth, which the prevailing trend suggests are the most important determinants in establishing a trusting relationship. When examining participants' perceptions of the manipulated variables revealed that perceived levels of anthropomorphism played a role in enhancing the positive effects of perceived competence and perceived warmth on attributed trustworthiness. This implies that constructive anthropomorphism plays an important role in making the propensity to trust robots stronger and more ecological, at least in the case of known determinants such as competence and warmth.

On the whole we then have a physical trust, which is largely dependent on folk physics, but which is regularly integrated with knowledge from experimental physics. When, instead of relying on the natural laws that govern the world, for achieving our purposes we leverage the psychological resources of other individuals, then we are using the intentional trust. Among the elements that constitute the second type of trust are included the reputation, the behavioral style, and the ethics and the social ranking of the individual we are considering whether to trust (Van den Brule et al. 2014; Woerdt and Haselager 2019).

A similar disposition to trust is *deference*. We are deferential, when we rely on a physician to diagnose and treat a disease, when we call a repairman to fix an appliance, or if we consult a vocabulary to find out what a particular word means. "Deference" today is a controversial term. While in individual life being deferential is an attitude that is often associated with impotence and lack of enterprise, from the point of view of social interactions it is an essential component of a constructive way of living together. Social deference, in fact, is not a mechanism of submission, but a smart strategy aimed at maximizing personal utility and building productive social relationships. It is the attitude that leads the individual to take advantage of the best confidence that someone else has with a certain area of knowledge or experience.

Likewise, to the distinction between physical trust and intentional trust, a similar distinction has to be made in the case of deference. The word "deference" refers to two different things. On the one hand, it alludes to the mechanism of dominance and submission. This is the field of the ethology of deference. It is about understanding how social hierarchies are established and how the biological constraints that are typical of a certain species influence how social and sexual hierarchies are shaped. From this perspective being deferential toward somebody is not the result of an intelligent strategy of social distribution of knowledge, but the result of domination of the fittest in the competition for the resources. In the animal kingdom, social hierarchies are variously distributed with regard to their flexibility. In some species, social rank is written into the DNA, in others it is the result of a sophisticated social negotiation. The same happens with male/female dominance in the different animal species. The dominant role may depend on the animal's size, the type of parental care required by the adaptive environment, or environmental changes themselves. It seems, however, that a slight asymmetry in sex roles ensures greater reproductive success (Jozifkova and Kolackova 2017). In any case, this kind of deference, which is characterized by being biologically grounded, is different from epistemic deference, which is a feature of the semantics of concepts and meanings.

In this respect, it refers to an epistemic mechanism that is constitutive of the functioning of concepts and meanings. Since nobody can know all meanings or can master all concepts, and since, nevertheless, it is necessary to use words and concepts whose content we do not govern in whole or in part, then we are forced to rely on the best confidence that others have about those same matters. Being deferential, in this second sense, has to do with cognitive economy, that is, being fast and frugal in performing the most common cognitive tasks. Epistemic deference is the component where intentional trust is most irreversible, since it involves knowledge whose availability is virtually inaccessible. Or, rather, the knowledge of others toward which we are normally deferential is regarded as too expensive. Perhaps it is a matter of becoming an engineer or physicist, or otherwise mastering broad and complicated areas of knowledge, engaging in an activity that is definitely too expensive, if the expertise of others can be easily used. When we let ourselves fall back based on the belief that others will keep us from falling ruinously to the ground, as in a well-known group game, we are *trusting* other people. When we take a pill that the doctor has prescribed for us to relieve a certain pain, we are *deferential* to the doctor and to the corpus of scientific knowledge that is expected to have motivated him or her in prescribing that given pill. Epistemic deference, then, consists of dependence on the knowledge of others to achieve one's own goals. It is a key component of trust, especially intentional trust, but should not be confused with it.

This might explain certain typical differences in placing trust or deference in certain devices or others. Consider a typical calculator and a humanoid robot, like Grace. Grace is a humanoid robot targeted at the healthcare market and designed to interact with the elderly and those isolated by the COVID-19 pandemic. She uses Artificial Intelligence to diagnose a patient and can speak English, Mandarin and Cantonese. Grace can simulate the action of more than 48 major facial muscles and has a comforting demeanor designed to look a little like *anime* characters, almost a fusion of Asian and Western traits. Usually, we have a deferential attitude toward calculators, which are typical "classical AI" devices. But, are we ready to have the same attitude towards a humanoid healthcare professional, like Grace? In the calculator case, it is not just a matter of trusting the calculator, but of simply being deferential to it. Perhaps we have a more deferential attitude towards classical AI artifacts and use preferably trusting humanoid robots.

But it is only a matter of degree and circumstances. Basically, the way we tend to adopt the attitude of trust or deference is the same whether it is directed toward our fellows or toward other animals and anthropomorphic machines. An experiment with oxytocin (de Visser et al. 2017) has shown that anthropomorphism in robots can increase trust in humans, and that the way we trust humanoid robots is generally consistent with the way we naturally regulate trust in human relationships. It is well known that exogenous administration of oxytocin increases the disposition to trust, to cooperate in group decision-making, and to engage in prosocial behavior in general. When oxytocin is administered to a group of humans and the anthropomorphic appearance of the artificial agents with which they are to perform tasks is modulated, it was founded that both exogenous hormone administration and anthropomorphizing correlate positively with the degree of trust in the artificial agents.

Another evidence for the claim that trust in humanoid robots is granted according to the same rules we follow in human interpersonal relationships comes from the propensity to obey an individual we deem as authoritative, even when what he or she tells us to do seems unreasonable. In such cases more than trust perhaps we should speak of deferential behavior. A well-known deferential attitude is that noted by Stanley Milgram's (1963, 1974) celebrated behavioral experiments. Milgram conducted an experiment in which participants were asked to administer increasingly severe electric shocks to another person if that person made a mistake in a learning task. Despite the potential harm, most participants followed instructions and delivered shocks up to 450 volts.

This experimental procedure has been replicated by comparing the level of obedience toward a human authority, i.e., a professor, and a robot issuing similar instructions to those received by the human experimenter. The result was that the level of obedience was very high in both situations, regardless of whether the source of authority was human or artificial (Grzyb et al. 2023).

5   To Conclude

A typical solution which is adopted by people seems to be what it could be called "selective deference." It is matter of balancing, according the above-mentioned "rule," trusting and deference taking into account all the robot, human and situation aspects. Selective deference is a type of deference that brings together the fear of losing control and the benefits of the social distribution of intellectual work. Selective deference is an epistemic attitude that hierarchizes, both implicitly and explicitly, the kinds of social knowledge to be most deferential toward (in which what the subject knows is more or less largely dependent on others) and those in which the subject instead has a direct commitment to the semantic content of what she is saying (or implies). Selective deference can work both implicitly and explicitly. However, the most important point is what rule we should follow to exercise selective deference. And the general rule is quite simple: Individuals (as rational agents) should be deferential to what they know is more or less dependent on others. And not should be deferential to that of which the subject has a direct commitment or expertise.

REFERENCES

Allen, C., Smith, I., Wallach, W. 2005. Artificial morality: Top-down, bottom-up, and hybrid approaches. *Ethics and Information Technology* (2005) 7:149—155 DOI https://doi.org/10.1007/s10676-006-0004-4.

Bruni D., Perconti P., Plebe A. 2018. Anti-anthropomorphism and Its Limits. *Front. Psychol.* 9:2205. doi: https://doi.org/10.3389/fpsyg.2018.02205.

Cangelosi, A., Schlesinger, M. 2015. *Developmental Robotics; From Babies to Robots*. Cambridge University Press, Cambridge (UK).

Christoforakos L., Gallucci A., Surmava-Große T., Ullrich D, Diefenbach S. 2021. Can Robots Earn Our Trust the Same Way Humans Do? A Systematic Exploration of Competence, Warmth, and Anthropomorphism as Determinants of Trust Development in HRI. *Front. Robot.* AI 8:640444. doi: https://doi.org/10.3389/frobt.2021.640444.

Carelman, J., 1925. *Catalogue d'objets introuvables et cependant indispensables*, Paris, Édition Balland, 1969.

De Visser, Ewart, Monfort, S., Goodyear, K. & Lu, L., O'Hara, M., Lee, M., Parasuraman, R., Krueger, F. 2017. A Little Anthropomorphism Goes a Long Way: Effects of Oxytocin on Trust, Compliance, and Team Performance With Automated Agents. *Human Factors*. 59. 116–133. https://doi.org/10.1177/0018720816687205.

Epley, N., Waytz, A., Cacioppo, J. 2007. On seeing human: A three-factor theory of anthropomorphism. *Psychological Review*, 114, 864–886.

Graziano M.S.A. 2013. *Consciousness and the Social Brain*. Oxford University Press, Oxford, UK.

Grzyb, T., Maj, K., Dolinski, D. 2023. *Obedience to robot. Humanoid robot as an experimenter in Milgram paradigm*, Computers in Human Behavior: Artificial Humans, Volume 1, Issue 2, 2023.

Jozifkova E., Kolackova, M. 2017. Sexual arousal by dominance and submission in relation to increased reproductive success in the general population. *Neuroendocrinol Lett* 2017; 38(5): 381–387.

Milgram, S. 1963. Behavioral study of obedience. *Journal of Abnormal Psychology*, 67, 371–378. https://doi.org/10.1515/pjbr-2019-0026.

Milgram, S. 1974. *Obedience to authority. An experimental view*. New York: Harper & Row.

Moore, G. E. 1925. *A Defence of Common Sense*, in "Contemporary British Philosophy" (2nd series), ed. J. H. Muirhead. Reprinted in G. E. Moore, *Philosophical Papers*, New York, Routledge, 2013.

Nof, S. Y. 2009. *Springer handbook of automation*. Berlin, Springer.

Plebe, A., & Perconti, P. 2022. *The Future of the Artificial Mind*. Boca Raton: CRC Press.

Van den Brule, R., Dotsch, R., Bijlstra, G., Wigboldus, D. H. J., Haselager, W.F.G. 2014. Do Robot Performance and Behavioral Style affect Human Trust? *Int. Journal of Social Robotics*, 6(4), 519–531.

van der Woerdt, S. & Haselager, W.F.G. 2019. When robots appear to have a mind: The human perception of machine agency and responsibility. *New Ideas in Psychology*, 54, 93–100. doi: https://doi.org/10.1016/j.newideapsych.2017.11.001.

CHAPTER 4

What Neurohacking Can Tell Us About the Mind: Cybercrime, Mind Upload and the Artificial Extended Mind

Paulo Alexandre e Castro

1   INTRODUCTION

In this chapter we seek to cross some elements of the philosophy of mind, namely, the myth of mind uploading and a certain perspective on the extended theory of mind (and not, as would be expected, with any computational theory of mind) with a specific type of cybercrime, neurohacking, to outline and present what we call the Artificial Extended Mind (AEMt) theory. Thus, in practical terms one can think, on the one hand that the extended mind theory is not just one theory among others but that it is something deeply connected to human reality and, on the other hand, that although the fanciful prospect of mind uploading has been generally accepted as such, alarming signs are beginning to be seen, precisely through neurocrime.

P. Alexandre e Castro (✉)
Institute for Philosophical Studies, University of Coimbra, Coimbra, Portugal

© The Author(s), under exclusive license to Springer Nature Switzerland AG 2024
P. Alexandre e Castro (ed.), *Challenges of the Technological Mind*,
New Directions in Philosophy and Cognitive Science,
https://doi.org/10.1007/978-3-031-55333-2_4

Philosophers of mind in particular, and theorists of transhumanism in general (from neuroethics to biotechnology), raise questions not only about the nature and conditions of mind but also post-humanism questions such as the cyborgization of the human or the future of human mind (for example, Bostrom, 2014). The general idea is that to be human means to transcend and, therefore, to transform. Implicit in this is also the idea that humans must adapt to this fast changes that are occurring in the human world, or if one prefers, like Craig Venter said (2015): "We're going to have to learn to adapt to the concept that we are a software-driven species and understand how it affects our lives. Change the software, you can change the species, [change] who we are."

In this multifaceted (and quite challenging) framework of hermeneutical possibilities, philosophy of mind is trying to provide some perspectives that reconcile the subject of the human mind with new emerging technologies such as Artificial Intelligence. However, the answers proposed by philosophy, which cover a varied spectrum of philosophical theories, do not always advance with the desired speed and even less so with the speed at which new questions arise. Even so, some contemporary philosophy, specifically the philosophy of cognition, the philosophy of technology and the philosophy of mind, are creating and crossing paths of reflection and interpretation with neurosciences, bioethics, law, criminology, computer engineering (etc.) that may contribute in the near future to a greater understanding of the human mind. Theories such as the Embodied Mind (EM), the Computational Theory of Mind (CTM), or the thesis of the Extended Mind (EMt), among others, may undergo significant reformulations or adaptations, and will certainly be more expressive for this understanding.

One of these questions is posed by the premise known as mind uploading—also known as Whole Brain Emulation (WBE)—, which is perhaps the most challenging question, not only for philosophers, but also for engineers, computer scientists, neuroscientists, deep down, it is a challenge for all areas of human knowledge.

This essay is divided into three main sections. In the first section, we will revisit some of the literature produced on mental loading and on the extended theory of mind (in his major figures). In the following section, neurocrime will be addressed, with the aim of envisioning possibilities for understanding the concept of mind and (consequently) brain in this specific scenario. From here, one seeks to understand how to structure a theory of the brain—mind relationship and the implications for a philosophy of mind. Finally, sketch a new configuration of the theory of the extended mind, which, according to the observations and premises raised by the previous reflections, will allow, on the one hand, to mention that downloading some specifics parts of the mind, it's already a reality and, on the other hand, allowing to show that an extended mind is or can only be understood as an artificial mind (AEMt). It is hoped, therefore, that this essay can challenge established paradigms, and, thus, create conditions for future reflections not only on the integration of artificial intelligence in other configurations of the human mind, but also on reality, which is also increasingly artificial.

2   Initial Questions About the Mind Upload and Extended Mind

The mind-upload question can be seen as a modernized version of the "brain in a vat," the thought experiment proposed by Gilbert Harman in 1973 (in fact, this was also derived from Descartes' exercise on the evil demon hypothesis) and basically, states that the mind can be transferred from a brain to any digital device (contents migrate from a human brain to an artificial environment) performed by some made-up brain—computer interface. Succinctly, it is said with this hypothesis that everything that can be in the mind is subject to a transfer, in the same way that a transfer of data between a smartphone and a computer is carried out.

In a sense, this hypothesis raises a set of observations (or problems) right away:

- the first observation to be made concerns the fact that the mind seems, at first glance, to present itself as a result of all brain or neuronal activity and, therefore, to claim a biological substrate or base without which it cannot exist;
- second, it calls into question everything that it is considered necessary for the formation of the mind as the intervention of the environment, the processes of socialization, etc., in a few words it presents a problem to the EMt (in general terms), which stresses the deep co-dependence of brain (and body) with the world (e.g., Sutton 2010; Menary 2010);
- third, and still in connection with the aforementioned, it puts any theory of the EM on an uncanny hold, as it seems possible to repli-

cate the mind on a computer—assuming here that the existence of some "affective computational architectures" have already been found (after the studies of Sloman and Croucher 1981, and Picard 1997)—and, therefore, there is no need for human bodies (unless they are considered producers of content to be transferred);

- fourth, and probably the most problematic observation: the mind uploading (MU) thesis seems to be, at the same time, dualistic and functionalist. Dualistic, because it treats the contents of the mind as something that can be moved from one side (brain) to another (computer), just as a file can be transferred from a flash drive to a computer (Hauskeller 2012, p. 196), and functionalist, because it determines that the mind can be both a function of brain processes as a function of digital processes.

Taking a quick tour throughout the literature produced around the theme, we find some curious positions (not strictly from the philosophy of mind). From Moravec (1988) to Kurzweil (2005), or to Wiley (2014), the idea seems to be that technology will be capable to transfer one's brain to some computer, which will provide an opportunity for humans to continue their existences (outside their biological natural bodies) through software machines. In the same line of thought, for Sandberg (2014), the mind upload technology would bring several benefits (e.g. allowing the space travel of contents instead of biological humans) and prevent human annihilation.

Hanson, in his book *Age of Em: Work, Love and Life when Robots Rule the Earth* (2016), describes a society built on uploading that could disruptively transform all forms of social structuring. One question that immediately arises, beyond these ideally fictionalized scenarios, is to know the real plausibility of this hypothesis.

Agar, reflecting on the problems found in the topic, argues that there is a sort of "unbridgeable gap," meaning, by this, that there is a difficulty between what human cognition is and what computer processes are: "saying that computers can actually think makes the same kind of mistake as saying that a computer programmed to simulate events inside of a volcano may actually erupt" (Agar 2011, p. 24).

Similarly, Corabi and Schneider (2014) also mention that it is not possible to admit this thesis, from the outset, because the whole argument is doubtful. They say that even if the simulation or replacement of the brain by a some digital-neuronal network may be conceivable (for instance, like a thought experiment), this does not mean that it is feasible (that become real). Arguing in this line of thought, for Piccinini, the MU hypothesis is not one that should be seriously considered (Piccinini 2021); nothing guarantees that this is possible because the mind, or, rather, each mind, holds (as far as we know) very particular states that include emotions, beliefs, personality traits, etc., and, therefore, hardly anything could be subject to a digital transfer (note, however, that Piccinini, in his last book, defends the computational theory of mind as the best theory for understanding the mind).

Differently, Clowes (2013, 2015) suggests that some digital technology devices (and even prostheses) are already examples of extended mind and consciousness (and, therefore, we would already be partially charged). For example, when looking at some (digital) memory systems, one can see that they expand and augment or even replace content that would otherwise be reserved for human biological memory (Clowes 2013). Thus, it is fair to say that some of the technologies that we are already using on a daily basis seem to be reshaping our memory, such as the well-known "google effect" in which users prefer to retain the path to information, but not the information itself (Sparrow et al. 2011; Wegner and Ward 2013).

To understand this problematic issue, as mentioned earlier, the EMt says that a mind is not just the brain. In fact, when Andy Clark and David Chalmers (1998a, b) formulate this initial thesis they already argue that the extended mind goes beyond the brain and body and extends to the world, to physical reality. It is easy to see that with such a thesis there seems to be no plausible connection with the MU. But one must keep in mind that the second wave of extended mind theory, which focuses on integration rather than replacement (of internal and external cognitive vehicles), may open a diferent perspective; as a matter of fact, it suggests that the boundaries of cognition are fluid and depend on the technological practices that we share. But we will return to this topic in section 4 of this chapter and see that there are other premises and circumstances in which this skeptical scenario may change.

Now, what can be said about mind uploading and how can this be considered an open door for neurocrime, for neurohacking activities? From a philosophical point of view, the possibilities and challenges that neurocrime opens up require a unique agenda.

3   SOME NOTES ON NEUROCRIME

The first published articles on neurocrime began to appear in 2001 in the renowned journal *Nature*, which in that year published two articles entitled, "Into the Mind of a Killer", and "A Danger to Society." In 2007 and 2010, *Nature* returned to this theme with "Abnormal Neurosciences: Scanning Psychopaths," and "Science in Court: Head Case," respectively. However, *Libération* in 2009 published "Un Juge Italien Découvre le Gène du Meurtre" and in the same year, the British newspaper *The Times* published "The Get Out of Jail Free Gene" (in November 2009). The BBC itself, in its online publication, reported on scientific advances around such a topic: "Psychopaths: Born Evil or With a Diseased Brain?" (November 2011). But it was in 2009 that an article appeared in *CRN News* entitled "Hacking the Mind: Why Your Brain Might be the Next Target." This led to the emergence of terms such as brainjacking or neurohacking. In ZDNET (2012), one could find the article "Mind Hackers Could Get Secrets from Your Brainwaves." A few years later, Forbes (2017) published "Hacking the Brain: The Future Computer Chips in Your Head."

The question to ask is: what is neurocrime? The concept of neurocrime it is not an easy one. Since its initial formulations, it has been confused with somewhat fanciful formulations anchored in the history of science and some ramblings of science fiction. There is also some disorientation (say, on the criminal and legal level) because, in fact, the border between cybercrime and neurocrime is blurred when the first constitutes, in most cases, the support and necessary condition for the second to occur. But the concept can configure other conditions and we will provide some examples of neurocrime (one of them is scientifically reported over 40 years ago that had nothing to do with cybercrime, at least in any of its now-known forms). That is to say that the concept of neurocrime can also be read as the result of an action (or several actions, procedures, or specific circumstances) in which occur (directly or indirectly) neuronal consequences for individuals. Putting it clearly: in our digital era (or, as Lipovetsky likes to call it, the hypermodern times), most of what is called neurocrime cannot exist without cybercrime.

Making a brief historical reference, the difficulty begins with the advancement of new areas of investigation in biology and the emergence of psychology (and the initial separation from philosophy). Some of the reflection was mainly focused on the search to understand and/or to know the criminal's genetics or mind, thus creating the idea and, let's say, the prejudice, that there could be a genetic predisposition to crime, reminding us of some theories born from phrenology with all the misunderstandings that this has caused, at both the social and criminal levels. Shape and size, among other characteristics, would a priori determine the presence of a criminal. Here, and by irony of history, it was with Cesare Lombroso (a Jew) around 1870, who, when analyzing the skull of a criminal Giuseppi Villella, realized a marked depression, which led him to be considered the father of criminology (and, according to some authors, the father of neurocriminology). His controversial theory of criminal anthropology, elucidated in *The Delinquent Man* (1876), had two key points: that crime originates due to deformities of the brain and that criminals would be an evolutionary throwback (that is, to the most primitive species). Thus, criminals could be identified based on physical characteristics, such as a large jaw and a sloping forehead. Based on these measurements, Lombroso created an evolutionary hierarchy, highlighting northern Italians and Jews at the top of the hierarchy, and southern Italians (such as the criminal Villella) at the bottom, along with Bolivians and Peruvians.

These beliefs, based in part on phrenological pseudoscientific theories about the shape and size of the human head, flourished and encouraged throughout Europe and some US states in the early twentieth century. We mention to the irony of history, and Lombroso's theory (who was a Jew) turned out to be socially and scientifically disastrous for the Jews themselves, as we know.

But if this seems too far away, perhaps we can also recall, as an example, the use of psychometric and anthropometric tests in South Africa to identify the Caucasian who was one of the sources that instigated apartheid. This "evidence" is typically introduced to support the statement that the defendant should be released from criminal responsibility because of an underlying brain defect that would lead him to commit a criminal act (*see* Raine, 2008; Senner et al., 2015).

Let us open a parenthesis here to hypothesize that, in this specific way, this is probably the first form of neurocrime. That is, this evidence was introduced as mitigating evidence at the sentencing stage of capital crimes trials. Of course, for those who advocate the use of these neuroscience findings in the courtroom, the introduction of these per se will enable judges and juries to draw more accurate conclusions about whether or not a defendant is responsible for his or her actions (Horgan, 2011). In practical terms, supporting a decision based on this supposed evidence can turn the judgment into a mere clinical finding devoid of any sense of justice, which also means a shake-up in theoretical discussions founded by neuroethics.

Some of the (main) reasons against the introduction of neuroscience in court are, for example, that these images are unreliable in terms of explaining or predicting behavior; the overestimation that can be conveyed by its colorful presentation and scientific appearance; the misunderstanding of the scientific discourse by judges and juries that leads to a conclusive translation of studies that do not always revert to the totality or fidelity of the study; and the tendency to generalize from particular cases (what is generally considered the "existence of legal precedent"). Neuroscience is increasingly entering the judicial sphere in order to understand and prevent crime, revolutionizing the understanding of what drives bad or criminal behavior, but that is a different question.

4   NEUROCRIME: NEUROHACKING AND MINDHACKING

The rapid development of biotechnology (and other important areas like genetics) and digital technology has also led to the rise of cybercrime. This trilogy (biotech, digital tech and cybercrime) that already seemed dangerous, led in turn to the emergence of biohackers and, within them, neurohackers. In this regard, two films can be recalled to illustrate them: the first, *Minority Report* (2002), an American science fiction film directed by Steven Spielberg (with Tom Cruise), based on the 1956 short story of the same name by Philip K. Dick, which recounts the story of a special "precrime" unit that arrests criminals based on the foreknowledge given by three psychics or mediums called "precogs." Of note, precrime is a term (coined by Dick) increasingly used in the academic literature to describe and criticize the tendency of criminal justice systems to focus on crimes not yet committed. The second one is the Christopher Nolan film *Inception* (2010), in which the character Dom Cobb (DiCaprio) is a thief who specializes in extracting information from the unconscious of his targets while they are dreaming.

Taking these two films as a reference and taking a leap into reality, it is seen that more and more interdisciplinary studies between law and neurosciences appear, and the topic of (neuro)prediction and (neuro)prevention has not been forgotten (it appears that fiction and reality merge).

A variety of studies are dedicated to studying the neural causes of violent behavior; it is a research topic that runs through journals of criminology, law, psychology, psychiatry, neuroscience (among others) and the studies have showed that neuroprevention and neuroprediction can be a way to stop antisocial and criminal behavior (Nadelhoffer and Sinnott-Armstrong 2012; Aharoni et al. 2013; Monahan and Skeem 2016; Adolphs et al. 2018; Poldrack et al. 2018), even assuming that in the conclusions it can be said that more research is needed, in a way that can be used (for example) in law (Bigenwald and Chambon 2019).

Some questions must be posed: what do we need to know about neuro-cyber-crime, and to what extent can it help to shed light on MU and EMt? What is the relationship between all of them?

The first notion is that the concept has developed and started to encompass the processes through which neuroscientific knowledge can be used as a starting point for committing crimes. Another important notion is that the focus has shifted from the criminal's mind to the victim's mind, as it is about understanding how neurological information can serve the illegitimate purposes of organized crime.

When talking about neurocrime, one is not in the realm of fiction, gaming or virtual reality. It must be said clearly that neurocrime is a reality and that therefore it must be approached in relation to its intent and in relation to its full potential; it is about accessing data that are no longer biometric, but neural.

It is already known that the strong increase in electronic devices connected to the internet in recent years, together with the use and manipulation of personal data, is already a reality and has led to the establishment of new criminal paradigms known as cybercrimes. If you want to provide in numerical terms an idea of the devices that were connected to the network and expand the data in a decade, you will have an idea of how such illegitimate acts can be made desirable: in 2011, there would be about 9 billion devices connected to the grid; by 2020, that number reaches approximately (an estimate) to 40 billion. In this special digital ecosystem, the estimate includes devices known as "brain–computer interfaces" (BCI) and "transcranial direct current stimulators" (tDCS), which means, firstly, a flow of neural information, and therefore, secondly, an unavoidable exposure of that information to cybercrime.[1] In fact, these devices turn out to be much more vulnerable than one might think. As Ienca and Haselager (2016) have pointed out about recent findings: "have shown that BCIs are potentially vulnerable to cybercriminality. This opens the prospect of 'neurocrime': extending the range of computer-crime to neural devices." The authors call this type of neurocrime brainhacking (the illicit access to and manipulation of neural information and computation). But what can be really said about this way of hacking neural information?

There are already several studies, but let's look at the studies carried out by Rosenfeld and Wong (2017) and Martinovic et al. (2012) with "brain–computer interfaces" (BCI). In those studies, it was possible to obtain not only several pieces of autobiographical information, but also the extraction of delicate information such as pin codes, credit card number, location and date of birth, known club memberships, etc.

Something fundamental should be underlined in this regard: the use of these devices has allowed to obtain sensitive information, which means that there has been a considerable leap here—both from the point of view of neurocrime and from the point of view of philosophy—which makes it possible to place the question about the validity and scope of both MU and EMt.

Another thing that cannot be forgotten and that as we have said elsewhere, it is important to bear in mind in the appreciation of these themes, that "the brain works as a whole, whether for the appreciation of art, to do some math or just to live the daily stress" (Castro 2021a, p. 129). The brain-mind that works together seems to end up corroborating the argument of MU and the existence of neurocrime, because, instead of weakening the possibility of MU (apparently it would be easier to circumscribe just one area of the brain, for example memory, to perform downloading bank card pin numbers), ends up reaffirming its potential. Neurohacking could become a trend if we think about the possibility of constructing "artificial minds" (as simulations of real minds). We must remember that

---

[1] Transcranial Direct-Current Stimulation (tDCS) is a portable, wearable brain stimulation technique that delivers a low electric current to the scalp. A fixed current between 1 and 2 mA is typically applied. tDCS works by applying a positive (anodal) or negative (cathodal) current via electrodes to an area. tDCS is a neuromodulation technique that produces immediate and lasting changes in brain function". See at: https://neuromodec.org/what-is-transcranial-direct-current-stimulation-tdcs

the well-known thesis of "brain emulation"—which seemed to be only and initially a mental experiment—poses the hypothesis of mapping the physical structure of the brain to the point of emulating mental states (copying or replicating as much as possible the small details of the structures of the human brain in such a way that the "answers" can be the same as the original). This possibility arises, we believe, supported by parallel projects that have been running for a few years, such as the "Blue Brain Project" or the "Human Connectome Project,"[2] which seek to map the human brain (neurons and their respective connections). One could argue that this is just science and academic research and that it has no impact on people's real lives; so, let us consider another example. One research project was created at MIT's Media Lab, by Kapur, called "AlterEgo" (2018). In short, AlterEgo is a prototype of a "gadget that sits on your face and lets you communicate silently with objects and other people. Kapur can use it to do things like change channels on his TV or order a pizza."[3] There are several examples of using technology to read the brain signals and capture/send information; for instance, the experience provided by "BrainNet," a "multi-person non-invasive direct brain-to-brain interface for collaborative problem solving" (Jiang et al. 2019), which enabled three people to directly communicate (or, as *Singularity Hub* wrote, "to directly transmit thoughts").[4] What should be highlighted in this section is that technology already exists that allows some forms of mind/brain reading. In fact, new discoveries are being made and new devices are appearing on the market (whether these are augmented reality glasses, brain chips, wireless headphones, gaming interfaces or brain stimulation devices) prevent or alert patients to health risks, to improve some kind of performance, or simply for fun, and yet the risk of exposure is increasing. This is a risk that can be taken advantage of by criminals and biohackers to

[2] See: https://www.epfl.ch/research/domains/bluebrain/, and https://www.human-connectome.org/

[3] "How it works AlterEgo: The device's electrodes track tiny electrical signals generated by your face and neck muscles when you silently read or talk to yourself. The signals are passed on wirelessly to a computer. The device also has bone-conduction headphones to give feedback and tell you (in a stilted, computerized voice) what anyone else who's also wearing AlterEgo is silently saying to you." Retrieved from the site: https://www.technologyreview.com/2018/06/19/142228/here-are-some-e-ways-to-upgrade-yourself-one-body-part-at-a-time/

[4] https://singularityhub.com/2018/10/09/how-brainnet-enabled-3-people-to-directly-transmit-thoughts/

access privileged information.[5] Add to this the fact that Artificial Intelligence is exploding and we have a perfect scenario (it is no coincidence that identity theft crimes are beginning to emerge through the use of AI).

5   THE EM, THE EMT AND THE AEM

The EM theorists are known for advocating the need for body structure to create a mind. They say that a what define a mind is strongly connected to the conditions of its bodily implementation (Varela et al. 1991; Shapiro 2011). However, it seems that the main argument against MU is based on a neuroreductionist assumption that actually interests them little or nothing (MU can focus, according to its purposes, on neuronal activity, if one prefers an attitude of neurocentrism that is compatible, for example, with the brain-in-a-vat hypothesis).

As seen, all proponents of the extended mind consider that human minds cannot be explained solely by what happens inside the braincase and therefore it is highly unlikely and reductive that a mind can be solely the result of brain-neuronal activity. They argue that the formation of the human mind requires all the processes and elements that are combined in the world (in a complex and dynamic way) and that contribute to human cognition, such as the brain, the body, the environment, the different structures of physical reality. In short, it is a mind that has phenomenological and functional specifications that are structurally determined by the biological, environmental, historic-sociological contingencies of its physical and bodily presence in the world.

In a sense, this means the existence of an experience-forming framework for the mind that demands the world and the objects and things of the world, such as connected digital objects. This issue will be taken up again later in the conclusion. It is not easy to formulate such hypotheses, and often an insurmountable debate seems to be established.

Massimiliano Cappuccio, in his remarkable paper "Mind-upload. The ultimate challenge to the embodied mind theory," draws attention to the

---

[5] For instance, see the following articles: "Neurotechnology and Artificial Intelligence: Where's the Limit?," *Open Mind - BBVA* (Feb. 2019). At: https://www.bbva.openmind.com/en/humanities/beliefs/neurotechnology-and-artificial-intelligence-wheres-the-limit/; "Taking Control of the Human Mind: How Close are Hackers to Brainjacking?", *The Cyber Express* (Nov. 2022) at: https://thecyberexpress.com/how-close-are-hackers-to-brainjacking/

need to recreate the categories/concepts with which we are familiar and, therefore, perhaps a different notion of freedom can be drawn from there. He observes that:

> If EM is right, and individuation is the price to pay in order to experience both the constraints and the freedom of real existence, then it is hard to believe that minds could ever migrate from their organic bodies to computers. But if we are not constitutively constrained by the existential cage in which we once fell, as MU ventilates, and our minds are only temporarily exiled from the transuranic world of pure essences, then maybe it is time to update our familiar categories for comprehending an unprecedented notion of freedom. (Cappuccio 2017, p. 446)

Despite the interest in this article, Cappucio did not consider other variables to get a complete picture. In fact, it wasn't just him; many continue to do philosophy without having their feet on the ground, without realizing the dimension of reality, of life. Returning to the theme, a few things still need to be said about this: the first one is that there are many incompatibilities between the different theories that have arisen and are fueled by the requirements they place (often metaphysical); the second is that there is also some incompatibility between what belongs to EM, EMt and MU is not always evident, for example the condition of exclusion of an embodied mind being downloaded, due to the notion of independence that exists in the mental subtract to be downloaded; and third, it seems to have a distance from the real world or, if you prefer, from the real experiences in which human beings live, which allow them to feed their theories on the one hand, and, on the other hand, to close themselves off to new possibilities. Thus, the dimension of reality should not (and cannot) be ignored. As already suggested, two things must be taken into account: the very way human beings live surrounded by protheses and objects, digital and connected objects, and the real existence of neurocrime (biohackers and neurohackers). These two pieces of evidence alone seem to add nothing to the discussion, but let's look at it in more detail.

The human world has always been a technological one. It's not our intention here to reproduce the narrative about *techné*: rather, we just want to say something quite simple: humans are technological beings (leaving aside the meaning that technological beings or technological society can have in the political and economic sense). Artificial Intelligence is no less complex than natural (or human) intelligence, and vice versa. Only in a technological set you can understand the essence of one another.

One should start by saying that, in human reality, we have been living with different types of prostheses for a long time. Prostheses that have been perfected in terms of both the materials used and their performance. So much so that the introduction of digital technology even changed the functionality of some of these prostheses. At the same time, the emergence of the internet and the development of digital technology (in recent decades), bringing with it the possibility of monitoring and measuring the functions of these prostheses in real time, changed the paradigm of the relationship between humans and technology, between humans and the world. Hauskeller was alert to this question and provided several examples:

> Artificial cardiac pacemakers implanted in the body have been successfully used for fifty years now, and artificial hearts and lungs are already being tested on live patients. Cochlear implants can replace ears and compensate for loss of hearing by stimulating nerve fibers directly in the brain. Similar devices are being developed for the restoration of sight. By connecting a video camera to a blind person's brain that sends signals directly into their visual cortex. (...) All these developments indicate that, in principle, our sense organs are dispensable since we can create their effects, or more precisely what they allow us to know and do, by other means. We don't even seem to need a body, at least not a functioning one, to interact with the world. (Hauskeller 2012, p. 188)

The author also advances with more extreme possibilities that reinvent, in a way and at the same time not only the "Ship of Theseus" (the thought experiment on whether an object that had had all its components replaced remains fundamentally the same object), but also the hypothesis of the brain-in-a-vat, posing the radical hypothesis of brain replacement. He says that "even the brain itself might be replaceable" (*iIbidem*). Alzheimer patients and others suffering from the effects of brain damage located in the hippocampus may soon be helped by having the damaged parts replaced by an artificial brain prosthesis, a microchip hippocampus (Berger and Glanzman 2005; Hauskeller 2012, p. 188). These technological advances might also provide an opening to MU, since it admits not only the interconnection between physiology and technology and at the same time between biological-neuronal functions and digital functions. It should be remembered, for the purpose of this chapter, that there is a close relationship that exists between some digital objects and human beings, such as "brain–computer interfaces" (BCI) and "transcranial direct current stimulators" (tDCS).

So, considering that humans have this close relationship to prostheses (they are even, in certain cases, essential for the maintenance of quality of life), that they relate to digital objects (BCI and tDCS), which are part of the world, one cannot fail to consider that such objects also enter the mind-forming equation; in other words and according to TMS fundamentals, they interfere with the formation of the mind. In fact, in this special digital ecosystem it seems to be easy to admit that they are a part of what it means to be human. And humans commit crimes, or in this specific case, neurocrimes.

Criminals want access to this neural information flow; they want access to the Extended Artificial Mind, which is nothing more than the relationship established between the human mind and connected digital devices. Everything that is part of EMt enters the AEM equation: all the digital objects that we saw before enter but also all the objects artificialized with "intelligence," that is, the inclusion of this new, but already fully integrated, reality of Artificial Intelligence (AI). Human intelligence makes worlds happen, inside and outside of reality, and, in this sense, its action is an extension of its intelligence. In fact, one cannot help but consider that AI is an extension of human intelligence (from the beginning, as it is born from it) and this is an integral part of AEMt. In a broad sense, all the AI-endowed objects must be considered an integral part of the extended mind. Thus, MU is not a possibility among possibilities but an already existing reality in his basic level (the existence of neurohackers and neurocrime validates such a reality) within the world that admits the validity of AEM (thus denying the existence of the MU myth). A short note: it must be said that the existence of neurocrime also means another concern for criminal agents and the law, since MU implies that individuals will be subject to the usurpation and manipulation of their identity (a common crime nowadays), but their personality, dignity and autonomy will also be called into question.

6   In Conclusion

We are aware of the difficulties that may affect the acceptance of AEMt. Other variables must certainly be considered in the defense and postulation of this thesis/theory and that deserve an agenda and a space for due discussion.

The truth of our mind still hides many mysteries, and it certainly involves factors that we have not yet considered or are not in a position to consider (perhaps our cognitive incapacity or lack of adequate language are some of the reasons, a kind of position closely related to that one of Colin McGuinn when he mentions that "it lacks the concept-forming cognitive procedures to fully grasp how mental properties such as the phenomenon of consciousness arise from their causal basis, meaning in simple terms, that humans do not have the right cognitive structures to capture such a phenomenon" (*Apud* Castro 2021b, p. 783).

However, AEM intends to assert itself as a serious possibility in this interpretative framework. Its connection with reality, with the materiality of the physical world and its objects, on the one hand, and, on the other hand, with the immateriality of information flows (data) and (some) mental contents (albeit rudimentary), allows express confidence in the future developments that may arise from this.

In conclusion and considering all that have been said, mind uploading is a strong theme that divides opinions (especially in the philosophical field). But for researchers in AI (from the neuromorphic computing to affect-aware devices) and bio(techno)logical hackers, there is enough reality in using simple devices to get some sort of neural/mind information.

Neurohacking done by "brain–computer interfaces" (BCI) and "transcranial direct current stimulators" (tDCS) is, in fact, a still weak and crude demonstration of mind down-or-up-loading, or, rather, a rude but real demonstration of the relationship between brain, mind and technology. In this sense, one cannot fail to consider the legitimacy of the AEM. The following must also be considered. From a philosophical point of view, neurohacking confirms a (certain) physicalist view of the mind (and, consequently, any form of dualism loses legitimacy). For philosophers of the EM and the EMt it may seem strange at first and difficult to accept this hypothesis, but when one thinks of the real existence of neurohackers (think about the illicit way of obtaining of data, neuronal information) as a download/upload action, that is, as an action only possible by admitting the existence of objects capable of performing such actions. It is legitimate to say that if you have an object/device to perform neurohacking this means that you must have the "object" hacked (be it a computer or a subject with brain and mental content) and therefore this is also an elementary way of confirming the Artificial Extended Mind thesis.

One can remember what Floridi (2014) said: we are *inforgs*. Humans are beings full of information, which means that in terms of AEMt, humans are potential victims of neurocrime—all of us—and that we can see our decisions conditioned or manipulated by criminals; in this sense, it is urgent to safeguard all sensitive information such as the autobiographical info that exists in our minds. What is important to underline is that the AEMt may be a possibility in future writings (about all these topics) about MU and EM.

So, admitting that,

1. there are some connected digital devices (like BCI and tDCS), that can be seen as examples of extended mind;
2. that these devices are part of the human reality (world);
3. that these devices can be/are used to extract sensitive information (as neuro information), configuring the existence of neurohacking and neurocrime;
4. that the existence of those illicit ways of obtaining neuro-information confirms the possibility of MU (even rudimentary);
5. that also from several studies it was possible to confirm and extract neural information;
6. That AI objects are an extension of human intelligence and also a form of EMt, which configures AEM, then,

it is possible to say that (in conclusion),

I. The mind uploading should not be consider a myth.
II. Neurocrime-neurohacking-brainjacking legitimate AEMt.
III. Reality is made up of all sorts of things that exist in the world (objects, devices that can whether be natural or artificial) and that are integrated in everyday life.
IV. The integration of different objects in our bodies/brains will increase and allow us to design new functions such as increasing memory and cognitive capacity (if we consider the functional integration of AI in the psychophysiological and cultural processes of the human mind).
V. Considering the previous premises, it is legitimate to consider,
   That an extended mind is an artificial (part of the) mind. Our technological mind is, in this sense, also an artificial (extended) mind. This is the main point of the Artificial Extended Mind (AEMt) theory.

It is known that Hauskeller's (2012) main concern is about the "hope of attaining 'digital immortality' through a completion of the ongoing cyborgization process" (note that himself mentions three objections: (1)

we don't know if a "perfectly accurate software emulation of a human brain will actually result in conscious experience," (2) "even that can be similar to conscious experience, we still have no guarantee that it will be anything like the experience of the mind," (3) but if this is the case, "it may be a different mind, or more precisely, a different self"). But despite this, he also argues that the self may be also preserved (in that cyborgization process) even with the danger that in the

> final step may prove one step too far and end the existence of the self. This is not only possible, but indeed very likely, since the final step is different from the previous ones in so far as it relies on the possibility of *copying* the self (instead of merely preserving it through a series of changes). (Hauskeller 2012, p. 200)

Our hope is that AEMt, above all, can be a way to rediscover the essence of what we are, of what it means to be a human being with a technological mind.

References

Adolphs, J. R., & Gläscher, Tranel, 2018. Searching for the neural causes of criminal behavior. *Proceedings of the National Academy of Sciences*, 115 (3): 451–452. https://doi.org/10.1073/pnas.1720442115

Agar, N. 2011. Ray Kurzweil and uploading: Just say no! *J. of Evolution and Technology 22* (1): 23–36.

Aharoni, E., et al., 2013. Neuroprediction of future rearrest. *Proceedings of the National Academy of Sciences*, 110 (15): 6223–6228. https://doi.org/10.1073/pnas.1219302110

Berger, T. W., and Glanzman, D. L. (eds.) [2005] *Toward Replacement Parts for the Brain: Implantable Biomimetic Electronics as Neural Prostheses*, Cambridge: MIT Press.

Bigenwald, A., & Chambon, V. 2019. Criminal Responsibility and Neuroscience: No Revolution Yet. *Frontiers in Psychology,* 10: 1–19. https://doi.org/10.3389/fpsyg.2019.01406

Bostrom, N. 2014. *Superintelligence: Paths, Dangers and Strategies.* Oxford: Oxford University Press.

Cappuccio, M. L. 2017. Mind-upload. The ultimate challenge to the embodied mind theory. *Phenom Cogn Sci* (2017) 16 (3): 425–448. DOI https://doi.org/10.1007/s11097-016-9464-0

Castro, Paulo Alexandre e, 2021a. Is there an Aesthetics Brain? A brief essay on the neuroaesthetics quantification of beauty. In: *Quantifying bodies and health. Interdisciplinary approaches,* Joaquim Braga and Simone Guidi (Eds). Coimbra: *IEF-eQVOLIBET-Universidade de Coimbra:* 127–138.

Castro, Paulo Alexandre e. 2021b. The paradox of "New Mysterianism" *vs* Esotericism. An Approach in Philosophy of Mind, *Philosophy Study*, 11 (10): 782–784. At: Doi:https://doi.org/10.17265/2159-5313/2021.10.006.

Chalmers, D. 2010. The singularity: a philosophical analysis. *Journal of Consciousness Studies*, 17: 7–65.

Andy Clark, David J Chalmers. January 1998a. The extended mind. *Analysis*. 58 (1): 7–19. doi:https://doi.org/10.1093/analys/58.1.7. JSTOR 3328150.; reprinted as: Andy Clark, David J Chalmers. 2010. Chapter 2: The extended mind. In: *The Extended Mind*. ed. Richard Menary. Cambridge: MIT Press, 27–42. ISBN 9780262014038.

Clark, A., & Chalmers, D. 1998b. The Extended Mind. *Analysis*, 58: 10–23.

Clowes, R. W. 2013. The Cognitive integration of E-memory. *Revue of Philosophy and Psychology* (4): 107–133.

Clowes, R. W. 2015. Thinking in the cloud: The cognitive Incorporation of Cloud-Based Technology. *Philosophy and Technology*, 282 (2): 261–296.

Corabi, J., & Schneider, S. 2014. If You Upload, Will You Survive?. *Intelligence Unbound: Future of Uploaded and Machine Minds*, Eds. Blackford, R, & Broderick, D. Wiley New York: Blackwell, 131–145.

Floridi, L. 2014. *The fourth revolution: How the infosphere is reshaping human reality*. Oxford: Oxford University Press.

Hanson, R. 2016. *Age of Em: Work, Love and Life when Robots Rule the Earth*. Oxford: Oxford University Press.

Hauskeller, M. 2012. My brain, my mind, and I: Some philosophical assumptions of minduploading. *International Journal of Machine Consciousness*, 4 (1): 187–200.

Horgan, John. 2011. Code rage: The "warrior gene" makes me mad! (Whether I have it or not). *Scientific American*.

Ienca, Marcello, & Haselager, Pim 2016. Hacking the brain: brain–computer interfacing technology and the ethics of neurosecurity. *Ethics and Information Technology*, 18 (2), doi:https://doi.org/10.1007/s10676-016-9398-9

Jiang, L., Stocco, A., Losey, D.M. et al. BrainNet: A Multi-Person Brain-to-Brain Interface for Direct Collaboration Between Brains. *Science Reports*, 9, 6115 (2019).

Kurzweil, R. 2005., *The singularity is near: When humans transcend biology*. New York: Penguin Press.

Martinovic, I., Davies, D., Frank, M., Perito, D., Ros, T., and Song, D. 2012. On the Feasibility of Side-Channel Attacks with Brain–Computer Interfaces. *Proceedings of the 21st USENIX Security Symposium. USENIX*.

Menary, R. A. 2010. Cognitive integration and the extended mind. *The extended mind*, ed. R. A. Menary. Cambridge: MIT Press: 227.

Monahan J. & Skccm J. L. 2016. Risk assessment in criminal sentencing. *Annu. Rev. Clinical Psychology*, 12: 489–513.

Moravec, H. 1988. *Mind children: The future of robot and human intelligence*. Cambridge: Harvard University Press.

Nadelhoffer T., Sinnott-Armstrong W. 2012. Neurolaw and neuroprediction: potential promises and perils. *Philos. Compass*. 7: 631–642.

Picard, R. 1997. *Affective Computing*. Cambridge: MIT Press.

Piccinini, Gualtiero. 2010. The mind as neural software? Understanding functionalism, computationalism, and computational functionalism. *Philosophy and Phenomenological Research*, 81 (2): 269–311.

Piccinini, Gualtiero. 2021. The myth of mind uploading, *In* Clowes, R. W., Gärtner, K, & Hipólito, I. (Eds) *The Mind-Technology Problem: Investigating Minds, Selves and 21ˢᵗ Century Artefacts*. Eds., I. Berlin: Springer, 125–144.

Poldrack, Russell A., Monahan, John, Imrey, Peter B., Reyna, Valerie, Raichle, Marcus E., Faigman, David, Buckholtz, Joshua W. 2018. Predicting Violent Behaviour: What Can Neuroscience Add?, *Trends in Cog. NeuroScience*, 22 (2): 111–123.

Raine, A. 2008. From genes to brain to antisocial behavior. *Curr Dir Psychol Sci*, 17: 323–328.

Rosenfeld, J. V., & Wong, Y. T. 2017. Neurobionics and the brain–computer interface: current applications and future horizons, *Medical Journal of Australia*, 206 (8): 363–368. https://doi.org/10.5694/mja2.2017.206.issue-8 https://doi.org/10.5694/mja16.01011

Sandberg, A. 2014. Ethics of brain emulations. *Journal of Experimental & Theoretical Artificial Intelligence*, 26: 439–457.

Sener, M. T., Ozcan, H., Sahingoz, S., & Ogul, H. 2015. Criminal Responsibility of the Frontal Lobe Syndrome. *Eurasian Journal Medicine*, 47 (3): 218–222. doi: https://doi.org/10.5152/eurasianjmed.2015.69

Shapiro, L. 2011. *Embodied cognition*. New York: Routledge.

Sloman, A. & Croucher, M. 1981. Why robots will have emotions. *Proceedings of the 7ᵗʰ International Joint Conference on AI*, 197–202. https://www.ijcai.org/Proceedings/81-1/Papers/039.pdf

Sparrow, B., Liu, J., & Wegner, D. M. 2011. Google Effects on Memory: Cognitive Consequences of Having Information at Our Fingertips. *Science*, 333 (6043): 776–778. https://doi.org/10.1126/science.1207745

Sutton, J. 2010. Exograms and interdisciplinarity: History, the extended mind and the civilizing process. In *The extended mind*. ed. R. Menary. Cambridge: MIT Press, 189–225.

Varela, F., Thompson, E., & Rosch, E. 1991. *The embodied mind. Cognitive science and human experience*. Cambridge: MIT Press.

Wiley, K. B. 2014. *A taxonomy and metaphysics of mind-uploading*. Los Angeles: Humanity+ Press and Alautun Press.

Wegner, D. M., & Ward, A. F. 2013. How Google Is Changing Your Brain. *Scientific American*, 309 (6): 58–61. https://doi.org/10.1038/scientificamerican1213-58

CHAPTER 5

On Neuroenhancement: Between Bioethics and Biotechnology

Adelaide Costa

1 Introduction

The concept of human enhancement has often been used vaguely, if not abusively, making it difficult to understand its application and its real scope. The non-consensual nature of the perspectives, defended by neuroscientists, philosophers, biologists and other researchers, in view of the questions that these "human changes" place on the horizon of those who think about them, is one of the main factors that hinder their understanding. But how can we not accept that there may be a disagreement about enhancement if there is not even a consensus on something as basic as normality? At least since Foucault, normality is no longer determined solely on the basis of difference or what was socially determined as such, therefore, today an even greater openness of mind is required in the face of the (bio)technological challenges we have at hand.

The issue regarding human enhancement is vast and delicate and requires clarification. There are many authors who have directly or

A. Costa (✉)
Porto, Portugal

© The Author(s), under exclusive license to Springer Nature Switzerland AG 2024
P. Alexandre e Castro (ed.), *Challenges of the Technological Mind*, New Directions in Philosophy and Cognitive Science, https://doi.org/10.1007/978-3-031-55333-2_5

indirectly addressed the problem, such as Eric Juengst (1998), Leon Kass (2003), John Harris (2007), Michael Bess (2010), Greely (2006), Allen Buchanan (2011), or Nick Bostrom (2008), but because we cannot cover them all, in this chapter, let us take only a few for the purpose of this essay.

Nick Bostrom and Rebecca Roache, like many of these authors, explains enhancement based on its denial, using an alleged distinction between enhancement and therapy, which ends up rarely being realized (and explains the reasons why):

> Enhancement is typically contraposed to therapy. In broad terms, therapy aims to fix something that has gone wrong, by curing specific diseases or injuries, while enhancement interventions aim to improve the state of an organism beyond its normal healthy state. However, the distinction between therapy and enhancement is problematic, for several reasons.
>
> First, we may note that the therapy-enhancement dichotomy does not map onto any corresponding dichotomy between standard-contemporary-medicine and medicine as-it-could-be-practised-in-the-future. [...]
>
> Second, it is unclear how to classify interventions that reduce the probability of disease and death. [...]
>
> Third, there is the question of how to define a normal healthy state. [...]
>
> Fourth, capacities vary continuously not only within a population but also within the lifespan of a single individual. [...]
>
> Fifth, we may wonder how "internal" an intervention has to be in order to count as an enhancement (or a therapy). [...]
>
> Sixth, even if we could define a concept of enhancement that captured some sort of unified phenomenon in the world, there is the problem of justifying the claim that the moral status of enhancements is different from that of other kinds of interventions that modify or increase human capacities to the same effect. (Bostrom and Roache 2008, 5).

From this quote we can see how the concept is linked to other areas and, therefore, to other very close concepts (which in itself explains the need for clarification). Note now the following quote from Greely, which introduces us to the notion of enhancement in the evolutionary context of Man:

> By definition, it seems like *enhancing* something will be a good thing. But when humanity is at stake, *enhancement* is controversial, even though human beings have historically struggled to enhance themselves. Humans used tools, fire, clothing and domesticated plants and animals to improve their safety, health, nutrition and power—making us "greater". The difference,

nowadays, is our growing knowledge of biology… The history of humanity is the history of *enhancement*. (Greely 2006, 87–88).

It should also be noted that, for Allen Buchanan, human and societal evolution arises as the result of historical enhancements: literacy and numeracy that enhanced the development of science; the agricultural revolution that boosted the development of cities and societies. The author considers that all these small revolutions had cognitive effects but not only that: they produced healthier societies and increased the standard of living (social effects); they increased productivity and thus improved the lives of all of us. (Buchanan 2011, 3–13).

An *enhancement* can never be understood without focusing on what seems essential to us: the reason and context of the intervention. Because any substance, method or technology, depending on the objective of its use, may be considered treatment or *enhancement* and, therefore, be subject to consensual acceptance or dubious embarrassment. Therefore, more than labeling technologies and interventions as good or bad, we must work on developing understanding and integration of techniques and applications, in order to enable the desired ethical judgment.

In our history, in our lives, there are achievements, evolutions and enhancements. Therefore, it is necessary to define which of these raise ethical questions and what answers we will have to these questions. It's just that it seems that everything that is biotechnological, biomedical, is bad; and that everything else is tolerable, good, or not even enhancement. This led to what Allen Buchanan called "biomedical enhancement exceptionalism" (Buchanan 2011, 11), the dogmatic assumption that because an enhancement involves biotechnology it is necessarily immoral. If we do not establish stricter criteria for differentiating the enhancement methods, we run the risk of obstinately freezing in our caves, when we could have discovered fire.

Let us reflect on a possible clinical case to help us understand the dimension of the problem and the understanding that can be obtained from it.

2   Clinical Vignette

J. is a 47-year-old man who has been in Psychiatry Consultation for around 3 years. He underwent treatment with antidepressant and anxiolytic drugs, after being diagnosed with a depressive episode. In the last 12 months he maintained exclusively psychotherapeutic follow-up, after scheduled medication withdrawal (given that he had a personality with anxious traits and an experiential context marked by high psycho-emotional distress).

In what was expected to be his "last appointment," given the clinical stability already sought, J. confronted the doctor with a request:

> Doctor, would it be possible to prescribe me Ritalin@? You know, my wife is unemployed, we have three young children, it's not easy. In addition to my job, I had to find a part-time job and that means I have to work a lot of hours.
>
> Lately I feel too tired to keep both jobs, but I have to. I know I'm not exactly sick, but I don't see any other solution. I have a friend who took these pills for a while and he told me that he felt like he was "superman": full of energy, in a good mood, he could work for hours on end without stopping! And I, at this stage, need that strength to continue. No one else can help my family and believe me, I've tried everything.

After the interview, and in the absence of a psychiatric diagnosis justifying the prescription of methylphenidate or any other drug, the doctor suggested behavioral procedures, aimed at reducing fatigue and improving general well-being. She informed about the absence of clinical indication for the requested prescription and warned about the adverse effects of such medication. She even proposed the collaboration of Social Services, in the sense of a socio-family intervention, with the aim of promoting external assistance to the family. The doctor shown willingness to continue monitoring J. in consultation (given the situation described) and proposed, if it were to be justified in the future, a possible pharmacological treatment. After the interview was over, she asked J. about any other matter she could help him with, to which J. responded negatively.

J. did not attend the subsequent appointment. A few months later, he was admitted to the emergency room due to a severe cardiac arrhythmia in the context of methylphenidate use. He was evaluated by cardiology and his situation was controlled, despite the high clinical risk.

He later confessed that he had purchased methylphenidate online, and that the medical team that assisted him, during the emergency situation, told him that it was very likely that he already had changes in his heart rhythm prior to taking the medication, and the situation worsened with the drug. As time "was always short," J. had not had medical examinations for several years, nor had he consulted his family doctor regularly.

This clinical case inspires clinical thinking and, above all, bioethical reflection. It raises numerous and important questions that deserve further discussion and which we will now focus on (at least, some of them).

1st Point: Treatment vs. enhancement

This is probably one of the least consensual topics, when considering the understanding that different authors have about neuroenhancement. It should be noted that as we cannot now go into the issue raised by the difficulty that psychiatric diagnosis often poses, especially when "less objective" diagnoses or less consensual situations such as personality disorders are at stake, it is interesting to reflect on the proposed topic.

Regarding the enhancement–therapy distinction, Norman Daniels proposed, in his book *Just Health Care* (1985), a model of "normal functioning," which seems to be based on John Rawls' *Theory of Justice* (2001) with the aim of justifying the administration of medical treatment only for those who present conditions that deviate from normality (that is, from the so-called "normal" human function).

Although his thinking allows for a potentially fair "rationing" of resources, it seems to present some limitations. One of them is the issue of normality, which we have mentioned previously, and which can present itself in a particularly problematic way in the area of Psychiatry. It is understood that the human being is also the reflection of a heterogeneity of traits and psychic characteristics that are in no way linear. For instance, Kass, who defends a more conservative position in relation to neuroenhancement (*Beyond Therapy—Biotechnology and the Pursuit of Happiness*), assumes the difficulty in differentiating treatment/enhancement (ambiguous interpretation), referring to Mental Health as a particularly problematic area.

For Daniels, everything that has to do with maintaining normal function falls within the scope of "treatment,, as opposed to enhancement (2009, 25–42). But we must put into question: in Psychiatry, and when we hear from our patients (that are taking antidepressants): "I have never felt so good, so happy in my entire life"; Is there not a certain enhancement "coupled" to the treatments? Or is it all just a certain "overvaluation" in patients' reports? Someone, with a personality with depressive traits, who at a certain point in their life suffers a depressive episode, cannot feel that, in fact, the antidepressant gives them skills that, prior to the depressive episode, they did not possess? A greater ability to socialize, for example. It doesn't seem impossible at all, or even unusual. Of course, an easy counter-argument could arise given the example we have just mentioned: to consider depressive personality traits as a pathology that requires pharmacological treatment. If so, it seems to us, that Psychiatry will have to be revisited; well Psychiatry and all medical art. Because the frontiers for diagnoses will necessarily have to be widened to a point where, we fear, extreme difficulty may arise in identifying "true" pathological states (ultimately, we fear, we run the risk of an excessive and inappropriate "psychiatrization" of society). Thus, we will have biometrically shorter children (according to the genetic potential inherited from their parents), who will be treated with growth hormone; people who are a little more obese, who will need treatment with methamphetamine derivates; more jealous or suspicious people, who will find themselves medicated with antipsychotics; less intelligent people, who will be coerced into taking psychostimulants...

These are some of the questions, but let's look at other examples: in the treatment of bipolar patients, they often express the desire to maintain a mood a little "above" the supposed "normality," especially because this gives them a greater sense of well-being. Isn't it also a form of enhancement? What about the current tendency to request that schoolchildren be given medication, whether with psychostimulants or antipsychotics, in order to standardize behavior? It becomes quite difficult, these days, to be what was once proudly described as a "child full of life."[1]

---

[1] Take the example of the USA: in 1970, around 150,000 children were medicated with methylphenidate (the drug having been synthesized in 1940 and marketed from 1960 onward). Data from 1993 estimated the "consumption" of methylphenidate in 2.6 million North Americans, the majority of them children between 5 and 12 years of age. The change in diagnostic criteria, with the appearance of the Diagnostic and Statistical Manual of Mental Disorders—DSM III, came to "neglect" the hyperactivity component, favoring attention deficit (so common in so many other pathological situations), so the number of diagnosed children was exponentiated (*See* Williams et al. 2004).

It seems to us, therefore, that in all therapies there may be an enhancement potential, which can be expressed in a more or less obvious way. This fact does not undermine the therapeutic objective, nor does it transform it into a threat.

Let us now consider another situation, which may seem to be outside the scope of the discussion on neuroenhancement, but which is justified taking into account the most recent biomedical developments and, therefore, technological developments involving the human mind. Let's take as an example a biotechnology under development (and already being tested), namely the "vaccine against Alzheimer's Disease" (Carrera et al., 2012).

For Daniels, vaccines are necessarily a treatment. We understand that this position is not the most appropriate; in fact, it seems to us that Preventive Medicine could be understood as a somewhat masked, but accepted form of enhancement, an enhancement that also ends up being reflected in a civic and personal duty: that of taking care of oneself.

It does not aim to treat diseases but rather to prevent them (and historically it has always been that way), that is, to enhance human capabilities in "combating" possible diseases. The Alzheimer's vaccine does not aim to cure any dementia, but rather to prevent it from occurring.

Let's think about the economic repercussions that come from treating this type of pathology: can preventing its appearance be considered an ethically valid and important enhancement for the person and also for society? In fact, and quoting John Harris: "the overwhelming moral imperative for both therapy and enhancement is to avoid causing harm and confer benefit" (Harris 2007, 58).

It seems to us that more than distinguishing what therapy is and what constitutes enhancement, it is important to evaluate cases individually, as already argued, taking into account the context and possible consequences. John Harris also tells us something that we consider very important and that allows us to conclude our reasoning: "therapeutics and enhancements are not exclusive categories" (Harris 2007, 56).

We think it is important to also explore some counter-arguments that have been put forward by authors who reject enhancement. Take, for instance, the "being healthy" as a criterion that contraindicates the use of intervention; and the "error" contained in the desire and search for a "better than good condition" (Kass 2003).

Therapy, in general, can be understood as something that corrects the dysfunction and enhancement can be described as something that goes beyond the so-called "normal" functioning (with all the weaknesses that this last concept seems to present).

Allen Buchanan gives us an example: altering the genes of an embryo to prevent a genetic disease will be therapy; altering the embryo's immune system to increase its defense capacity against infections will already be enhancement (as it goes "beyond therapy" (Cf. Buchanan 2011).

However, and similarly to what has already been argued in relation to Preventive Medicine, it seems that both examples can be understood (let us say, at the same time) as treatment and enhancement, albeit to apparently different degrees (in both, we are preventing diseases, whether genetic or infectious, but perhaps in the latter case we are conferring even greater "power" on the boosted embryo).

Having or not having health are notions that denote great fragility (which comes, for example, from the different perceptions of therapists and patients).

Here we enter a completely different domain that constitutes a profound mystery: what happens in the brain and what happens in the domain of subjectivity. In other words, the information that reaches the brain as a result of our conscious experiences is not always explicit as a causal link. Let's see, for example: when a person is depressed, only they can say (with their characteristics and individuality) what they feel. Or, rather, only she knows what she feels.

The doctor has the empathetic role of trying to put himself in her place (which we believe will never happen completely) and help her. We do not argue that there are depressions or "sick people": there are "people who are sick" or, following this line of reasoning, "depressed people."

Therefore, "being healthy" can also be understood as a subjective state, which minimizes the success in distinguishing between treatment and enhancement (possibly for J., the life situation in which he found himself and the physical and psycho-emotional exhaustion, would be incompatible with the notion of a total "health," even if the objective diagnostic criteria were not met, which often fail to disregard individual subjective experience).

Michael Bess also alerts us to another issue (cf. 2010, 641–655): the concept of health itself varies from culture to culture, from era to era, from a negative concept to a positive concept, which reinforces the idea of that we don't always have a solid basis to try to distinguish treatments from enhancements.

Leon Kass states that restoring health is good, but seeking a "better than good" condition is wrong. The question arises: is the doctor such a powerful being that he can determine, without a doubt, what is best for the patient? In an era where patient autonomy and participation in therapeutic decisions are prioritized, isn't this view excessively paternalistic and reductive of their freedom?

On the other hand, and understanding enhancement as something that goes beyond the disease, do we have reliable standards about what is better or worse in relation to our choices? We feel that this could lead us to unacceptable practices, which some authors warn us about in the face of the "slippery-slope argument." This argument, proposed by Schauer in 1985, proposes a kind of "slippery slope," intending to alert us to the fact that a particular act, apparently "innocent," when taken in isolation, can lead to a future set of events of increasing harm (Kamp 2002, 1–4); this will justify not making small concessions or interventions, apparently without major consequences, in the face of controversial situations.

The great bioethical challenge, in understanding enhancement, is the very definition of "limit," which differentiates between treatment and enhancement or, in other words, the limit beyond which a treatment can be seen as enhancement.

Despite the practicality of a clear differentiation, this is not always possible, so a careful assessment of each case must be carried out. It seems to us that the use of a systematic evaluation, such as the one proposed by Michael Bess (Cf. Bess 2010, 641–655), dividing enhancement into degree, mode and relative effect may be useful, especially because such analysis can extend to cases in which there is some uncertainty regarding the classification of the phenomenon as therapy or enhancement, facilitating the ethical assessment of situations.

Taking J.'s clinical case as an example, we would have different ways of improving his professional performance (the objective he intended): whether through behavioral procedures or the use of methylphenidate. In terms of degree, this would translate into a modest or more noticeable enhancement.

The relative effect of the action, highlighting the competitive advantage that comes from using one method or another, could lead us to think about a certain injustice, aggravated by the greater effect of the drug on cognition, when we understand that a J. pharmacologically enhanced would be much more capable of performing work tasks than non-medicated co-workers, leading to consequent repercussions in terms of personal gains (the company where J. works could even lay off other workers if J. were still able to perform more tasks).

Given the case of J., it seems to us not to be abusive to consider that this is an attempt of enhancement, especially because the user himself denies feeling "sick" and states that what he wants is to improve his capabilities that will give him greater profitability at labor.

However, it is not always so easy or consensual to reach this conclusion, which is why we understand that all cases that raise doubts must be evaluated in detail and with the utmost caution, taking into account the proposed distinctions. Having said that, let us now consider another issue that deserves careful analysis.

2nd Point: Efficacy vs. safety

Considering J.'s clinical case, we could assume that the medical attitude of not prescribing methylphenidate was due to two main factors: the absence of clinical indication (prescribing the drug would be going, as Kass tells us, "beyond treatment"), and the consideration of the physical risks associated with the use of methylphenidate (Wang et al. 1994; 2007).

In fact, if adverse effects are taken into account when prescribing a treatment, which may limit it (either due to the patient's refusal or due to the medical cost–benefit consideration), they become even more relevant if we consider an intervention for elective purposes as to improve cognition and performance, in the absence of disease.

Long-term adverse physical effects (15 to 20 years) in children with ADHD who have taken methylphenidate for long periods (from childhood to early adolescence) appear to be minimal (Cf. Jacobvitz et al. 1990, 677–688).

However, the physical effects that may arise from use after adolescence and during adulthood are unknown (Diller 2010, 50). In relation to the immediate effects, such as reduced appetite, agitation with insomnia, among other possible effects, they seem to be better tolerated by children than by adults with ADHD, especially because, as a rule, children take medication only for shorter periods, and adults often tend to self-medicate daily, as they feel they need the drug to be able, for example, to concentrate on academic or work tasks.

Another issue refers to the possibility of abuse and even dependence (it should not be forgotten that methylphenidate, despite having been announced as a safe and harmless drug, is structurally similar to the "precursors" d-amphetamines), which even though apparently absent in medicated children with ADHD, it has already been seen in adolescents[2] and adults (Morton et al. 2000, 159–164; Ozaki and Wada 2006, 89–99).

In this regard, we think it is relevant to focus on an aspect that is observed in some adult patients who seek psychiatric consultation in order to obtain a prescription for methylphenidate (some with a previous diagnosis of ADHD, others not). Those who do not meet criteria for ADHD often present evident impulsivity and, sometimes, a somewhat "sensation-seeking" profile (Roberti 2004), some even presenting a history of previous abuse of other substances, such as cocaine (this type of traits and conduct is often associated with both the consumption of illicit drugs and antisocial behavior).

In physical (physiological) terms, some evidence has already emerged about the continued and prolonged, and potentially negative, use of methylphenidate.

Thus, a decrease in cerebral vascularization was observed using PET (positron emission tomography) in up to 30% of people receiving medication, related to the impact that the drug has on dopamine levels (Wang et al. 1994), metabolic changes with increased brain energy consumption (Porrino and Lucignani 1987), atrophy or loss of brain tissue (Nasrallah et al. 1986), increased sensitivity to cocaine (Lambert 1999) and greater prevalence of anxiety and depression.[3]

An important aspect that any doctor must take into account when considering the administration of a drug is the "associated risk." In relation to J.'s clinical case, this risk was not properly explored, especially because there was a refusal to prescribe medication.

However, given what happened (the off-label purchase of the drug was not a situation considered by the medical team), wouldn't it perhaps be more prudent to anticipate similar situations, even requesting J a cardiovascular exam?

In fact, access to purchasing medicines via the internet is increasingly easier and more frequent, and it is also not uncommon to obtain

---

[2] Linda D. Marsh, Janice D. Key, and Tricia P. Payne, "Methylphenidate Misuse in Substance Abusing Adolescents", *Journal of Child & Adolescent Substance Abuse* 9:3 (2000): 1–14; Robert J. Williams, et al., "Methylphenidate and Dextroamphetamine Abuse in Substance-Abusing Adolescents", *American Journal on Addictions* 13:4 (2004): 381–389.

[3] National Institutes of Health (NIH), 2003, http://www.nih.gov/news/pr/dec2003/nida-08.htm

medicines from patients who usually use them, namely friends or family (Maher 2008, 674–675).

And this is a situation that needs to be considered, as far as possible, when evaluating our patients. It should be noted that for an individual with no known medical history, the study of some more important physical variables will certainly not be time-consuming or expensive (a simple electrocardiogram could have detected an arrhythmia, which would most likely discourage J. from taking methylphenidate).

Still in this regard, we think it is important to consider an article by Alexandra Ossola, in which the author warned that "the fake pharmaceutical industry is growing and we cannot do anything about it" (Ossola 2015, 29).

This chapter warns about the growing commercialization, especially via the internet, of drugs of dubious production, from antimalarials to vaccines, antibiotics, retrovirals, sildenafil citrate (Viagra®), among countless others.

Networks of drug counterfeiters tend to grow, and the chapter warns of the fact that around 90% of drugs purchased online come from countries of origin other than the country of purchase. Legislation and policing in relation to this type of trade are still clearly insufficient, meaning that high risks appear to arise for the health of the population who acquire their medical products through this route.

Another issue that we would like to highlight, and which in a way implies a "risk," is the fact that, by enhancing certain brain areas (for example with a drug), we can improve some cognitive functions but simultaneously worsen the functioning of other brain regions (this happens through the activation of neurotransmission pathways in different brain areas).

These facts suggest that the change in neurotransmission induced by the drug may improve brain function in certain people and have no effect, or even worsen brain function in other people who are already at an optimal level of functioning (Sahakian and Morein-Zamir 2011, 197–204).

In fact, the "Principle of the Inverted U-Curve," which governs the response, for example, to psychostimulants, tells us that cognitive performance is optimal for an equally optimal level of catecholamines (deficit or excess prove to be harmful in terms of operating mode). On the other hand, cognitive improvement with psychostimulants also depends on the individual's basal state, with a greater impact on cognitively "less gifted" individuals and even a reduction in performance in individuals who have a cognitive basal level very high (Micoulaud-Franchi et al. 2012). The response to psychostimulants will also depend on other variables, such as genetic susceptibility (ibidem.). Despite much that we already know about the functioning of the human body and the available pharmacological products, more or less expected adverse effects can always arise.

The decision regarding the prescription of a treatment involves an assessment of the benefit, given the expected adverse effects, and the sharing of sufficient information for adequate consent. If, in relation to treatments, the inherent risks can be a problem, in relation to the use of substances for enhancement purposes the problem (which involves an even greater lack of knowledge) seems to be more serious. A prudent attitude will be the most advisable, ensuring that *primum non nocere* (from Hippocrates).

3rd point: The value of human action

The risk that neuroenhancement may pose to the perception of ourselves is something that deserves due attention. Being unable to recognize the merit for what we have managed to do throughout our lives could be a reason for a life devoid of objectives, but also for a possible emptiness, when it comes to contemplating the action and value of others.

The article "Pool results: look who's doping" published in *Nature*, presented surprising results, revealing a high percentage of consumers of psychostimulants for the purpose of neuroenhancement (Maher 2008, 674–675).

However, the surprise effect tends to disappear, if we take into account the opinion of several neuroscientists, such as Michael Gazzaniga who highlights: "If we can improve our abilities to compensate for what mother nature did not give us, why is that so bad?" (Burne 2007).

Barbara Sahakian also stated the need to debate the issue of the use of this type of drugs, especially because, she states, "their use is already common not only among students but also among senior academics" (ibidem.).

And we don't think it's a culturally localized situation, as we tend to observe it in our country as well. We must therefore question the place of merit and the value of personal effort.

This is an issue that stands out in the studies of Michael Sandel, who fears the creation of an excessively powerful human action, new "Prometheans," who aspire to remake nature and Man, in order to satisfy their personal desires (Sandel 2002). This will, the author argues, reduce the value of personal effort and eliminate humility.

But let's go back to J.'s clinical case: a first reaction will be to consider that he is doing a kind of "cheating," which takes away the value of the action; that is, ethically it would be more acceptable, and even admirable, for J. to use only his "natural" resources to carry out his work tasks.

One would certainly praise his personal effort and the merit of his work. However, allow us to make one note that seems particularly important: J. requests the prescription of methylphenidate to "be able to support" two jobs, in order to provide for his family and J. assumes that he cannot do it without "help external."

It doesn't seem to us that J.'s "cheating" is necessarily done for personal gain and, in a way, his attitude even involves some altruism and solidarity. J.'s objective could be considered ethical, an ultimate search for the good, well-being and happiness of his family; however, the means may not be.

And his option would certainly be highly criticized by defenders of "fair means" and "self-control," as clearly J. approaches an extreme and loses rational control over the emotions that lead him to decide to neuropotentiate himself. evidencing, perhaps, a "temperance" below that which Aristotle recommends to the Virtuous Man (Aristotle, *NE*, Book III, 1117b23-30).

Of course, other analyzes can be done on the same situation. For example, from another bioethical-philosophical perspective, namely one of Kantian inspiration: J. would be reducing himself to a means and not an "end in itself," so his attitude would be ethically reprehensible, given the obvious instrumentalization of Man (Kant, 2002, 69).

The most important thing in all these issues is the ability to understand the concept in different contexts and the importance of its uses taking into account what it means to be human in a world increasingly governed by technological advancement.

3   In Conclusion

If we evaluate the situation from a utilitarian bioethical perspective, perhaps we will find ethical validation for the action. In this case, enabling less pain and greater happiness for others other than J. (namely his family), could correspond to a good action.[4] It is felt that the utilitarian view is the one that predominates in the thinking of those who defend neuroenhancement as a valid and ethical practice, minimizing concepts such as the "value of human action," given the good consequences that may arise from enhancement interventions, for the maximum number of people that these methods can reach.

It is clear that the analysis needs to be sufficiently broad and integrated not only in the academic, scientific and clinical context but also in the individual and social context. Especially because, we emphasize, justice and equality in access to biotechnologies prove to be priority issues when defending neuroenhancement as a practice that we can/should "accept." The questions that bioethics raises reflect the potential that biotechnology presents for a rapidly changing society. A technological society that must be prepared for greater and more accelerated changes. In this sense, neuroenhancement should appear as an educational topic on scientific agendas, but also on political, economic and social agendas. The society of the future will require, more than ever, an education for the technological mind.

REFERENCES

Bess, Michael. 2010. Enhanced Human versus Normal People: Elusive Definitions. *Journal of Medicine and Philosophy* 35: 641–655.

Bostrom, Nick, Roache, Rebecca. 2008. Ethical Issues in Human Enhancement, *In* Jasper Ryberg and Thomas S. Peterson (Eds). *New Waves in Applied Ethics*. Hampshire: Clark Wolf, 2008, 1–27.

Buchanan, Allen. 2011. *Better than Human*. Oxford: Oxford University Press.

---

[4] In the same way that the individual aspires, by nature, to their individual happiness, promoting the well-being of everyone (in this case we make the analogy to J.'s family) will be a good action. This utilitarian or consequentialist ethics argues that a person must articulate their personal interests with common interests, providing maximum utility to everyone involved. One of the principles of utilitarianism, that of "Maximum Happiness," defends that actions are fair to the extent that they produce happiness, creating a kind of "calculation" of well-being, evaluating predictable advantages and disadvantages in relation to the possibilities of action . For a deeper understanding of the utilitarian current, see: John Stuart Mill, *El utilitarismo: un sistema de la lógica*, Libro VI, capítulo XX (Madrid: Alianza Editorial, 2014). If one prefers, see a modified perspective in: Peter Singer, *Ética Prática* (Lisboa: Gradiva, 2002).

Burne, Jerome, 2007. Can a pill really make you brainy?, *MailOnline*. Available at: http://www.dailymail.co.uk/health/article-504507/Can-pill-REALLY-make-brainy.html.

Carrera, Iván *et al.*, 2012. Vaccine Development to Treat Alzheimer's Disease Neuropathology in APP/PS1 Transgenic Mice, *International Journal of Alzheimer's Disease*. http://dx.doi.org/https://doi.org/10.1155/2012/376138.

Daniels, Norman. 2009. Can anyone really be talking about ethically modifying human nature?, *In* Savulescu, Julian, Bostrom, Nick (Eds), *Human Enhancement*, Oxford: Oxford University Press, 25–42.

Daniels, Norman. 1985. *Just Health Care*. Cambridge: Cambridge University Press.

Diller, Lawrence, H. 2010. The Run on Ritalin: Attention Deficit Disorder and Stimulant Treatment in the 1990s, *In* Martha J. Farah (Ed.), *Neuroethics—An Introduction with Readings*, The MIT Press.

Greely, Henry T. 2006. Regulating Human Biological Enhancements: Questionable Justifications and International Complications. *Santa Clara J. Int'l L.* 87: 87–109.

Harris, John. *Enhancing Evolution—The Ethical Case for Making Better People*. EUA: Princeton University Press, 2007.

Jacobvitz, D. *et al.*, 1990. Treatment of attentional and hyperactivity problems in children with sympathomimetic drugs: A comprehensive review, *Journal of the American Academy of Child and Adolescence Psychiatry* 29: 677–688.

Juengst, Eric. 1998. What does enhancement mean?. *In*, Erik Parens, ed. *Enhancing Human Traits: Ethical and Social Implications*. Washington DC: Georgetown University.

Kamp, George. 2002. Slippery Slope Arguments in Bioethical Debates, *Europäische Akademie* 31: 1–4.

Kass, Leon R. 2003. Ageless Bodies, Happy Souls, *The New Atlantis* 1: 9–28.

Lambert, N. 1999. Ritalin and its Cousins: Rx or Gateway Drugs?, The Regents of the University of California 27:34. At: http://www.berkeley.edu/news/berkeleyan/1999/0512/ritalin.html.

Maher, B. 2008. Pool results: look who's doping,. *Nature* 452: 674–675.

Marsh, Linda D., Key, Janice D., Payne, Tricia P. 2000. Methylphenidate Misuse in Substance Abusing Adolescents, *Journal of Child & Adolescent Substance Abuse* 9, 3: 1–14,

Micoulaud-Franchi, J.A. J. Vion-Dury, and C. Lancon, 2012. Neuroenhancement in healthy subject? A French case study», *Thérapie*, 67 (3): 213–21.

Mill, John Stuart. 2014. *El utilitarismo: un sistema de la lógica*, Libro VI. Madrid: Alianza Editorial.

Morton, W. Alexander, D. Pharm, B.C.P.P., Gwendolyn G. Stockton, *et al.* 2000. Metylphenidate Abuse and Psychiatric side effects, *Primary Care Companion J Clin Psychiatry* 2 (5): 159–164.

Nasrallah, H. *et al.*, 1986. Cortical atrophy in young adults with a history of hyperactivity in childhood, *Psychiatry Research* 17: 241–246.

Ossola, Alexandra. 2015. The Fake Drug Industry is Exploding, and We Can't Do Anything About It, *Newsweek—Phony Medicine*: 28–35

Ozaki, S., and K. Wada. 2006. Characteristics of methylphenidate dependence syndrome in psychiatric hospital settings, *Nihon Arukoru Yakubutsu Igakkai Zasshi* 4 (2): 89–99.

Porrino, L.J., Lucignani, G. 1987. Different patterns of local brain energy metabolism associated with high and low doses of methylphenidate, *Biological Psychiatry*, 22 (126): 126–138.

Rawls, John. 2001. *Uma Teoria da Justiça*, trad. Carlos Pinto Correia. Lisboa: Editorial Presença.

Roberti, Jonathan W. 2004. Review of behavioral and biological correlates of sensation seeking», *Journal of Research in Personality* 38: 256–279.

Sahakian, Barbara J., and Sharon Morein-Zamir, 2011, «Neuroethical issues in cognitive enhancement». *Journal of Psychopharmacology* 25(197): 197–204.

Sandel, Michael J. 2002. What's Wrong with Enhancement?, *The President's Council on Bioethics*. At: https://bioethicsarchive.georgetownedu/pcbe/background/sandelpaper.html0.

Singer, Peter. 2002. *Ética Prática*, trad. Álvaro Augusto Fernandes. Lisboa: Gradiva.

Wang, Y., Y. Zheng, Y. Du, D.H. Song, *et al.* 2007. Atomoxetine versus methylphenidate in paediatric outpatients with attention deficit hyperactivity disorder: a randomized, double-blind comparison trial, *Aust NZ J Psychiatry* 41(3): 222–230.

Wang, G. J., N. D. Volkow, J. S. Fowler, R. Ferrieri, *et al.* 1994. Methylphenidate decreases regional cerebral blood flow in normal human subjects», *Life Science* 54. Available at: http://science.naturalnews.com/pubmed/8114609.html.

Williams, Robert J. *et al.*, 2004. Methylphenidate and Dextroamphetamine Abuse in Substance-Abusing Adolescents, *American Journal on Addictions* 13, 4: 381–389.

PART II

Challenging Mind: Between Cognition and Art

CHAPTER 6

Consciousness, Theory, and Mental Appearance

David Rosenthal

1   INTRODUCTION

I contrast one-factor views of consciousness with two-factor views. On two-factor views a state's being conscious consists in something distinct from the state itself; one-factor views deny that. Examples of one-factor views include the first-order theory of Fred Dretske, the first-order approach of Thomas Nagel, and Ned Block's conception of phenomenal consciousness. Examples of two-factor views are a higher-order theory of consciousness, such as what I have defended elsewhere, and the global-workspace theories of Bernard Baars and of Stanislas Dehaene and Lionel Naccache.

I argue that one-factor views preclude not only a useful explanation of consciousness but also any informative description of what it is for a mental state to be conscious. Because of that and other related factors, one-factor views are strongly anti-theoretical. I consider several indicators of

D. Rosenthal (✉)
Graduate Center, City University of New York, New York, NY, USA

© The Author(s), under exclusive license to Springer Nature
Switzerland AG 2024
P. Alexandre e Castro (ed.), *Challenges of the Technological Mind*,
New Directions in Philosophy and Cognitive Science,
https://doi.org/10.1007/978-3-031-55333-2_6

this anti-theoretical attitude, including advocacy of a hard problem or explanatory gap and the appeal to allegedly pretheoretic intuitions.

Some have found one-factor views appealing because they reject any coherent contrast between the mental appearance and the mental reality of conscious states. But denying that distinction derives from a misunderstanding of our commonsense conception of consciousness. And independently of that it is also theoretically indefensible. Examining the rejection of a distinction between the mental appearance and mental reality of conscious states helps us appreciate the shortcomings of a one-factor view.

2   TWO APPROACHES TO CONSCIOUSNESS

The current literature about consciousness often taxonomizes views into first-order (FO) and higher-order (HO) views. But a more revealing division, I'll argue, is more general than that, and relies on distinguishing between what I'll call one-factor (1F) and two-factor (2F) views.

1F views hold, with Thomas Nagel (1974), that the property of a mental state's being conscious is intrinsic to that state. By contrast, higher-order (HO) and global-workspace theories (GWT) are 2F views, which hold that a state's being conscious is a matter of some factor distinct from the state itself.

HO theories all explain what it is for a state to be conscious by appeal to one's being aware of that state. FO theories deny that any such HO awareness (HOA) figures in a state's being conscious. The views of Nagel, Fred Dretske, and many others are FO: Being conscious is intrinsic to the state.

On GWT, by contrast, a state is conscious if its content is available for global downstream effects. GWT is also a FO view, since it does not rely on any HOA (Baars 1997; Dehaene and Naccache 2001). But both HO and GWTs explain consciousness by appeal to a factor that's distinct from the state, a HOA or global availability for downstream processing. Indeed, it is natural to fold the basic idea of HO theories into a GW theory, as Lionel Naccache arguably does (2018). Both HO and GW theories operate in similar ways, despite their being taxonomized differently on a distinction between FO and HO theories.

This distinction between 1F and 2F theories parallels the distinction sometimes drawn between local and nonlocal theories of consciousness (e.g., Michel and Doerig 2022). On local theories so described, the neural correlate of a perceptual state's being conscious is local to the relevant area of perceptual cortex. On a nonlocal theory, the neural correlate is elsewhere in cortex, typically in prefrontal cortex.

As I'll argue shortly, however, it's important to cast theories of consciousness in distinctively psychological terms, since consciousness itself is a distinctively psychological phenomenon. The distinction between local and nonlocal theories is cast not in psychological terms, but by appeal to underlying neurological correlates. By contrast, both 1F and 2F theories are cast in distinctively psychological terms. So, in what follows I'll stick with that distinction.

Most contemporary discussions of consciousness tend to focus on perceptual consciousness. But just as the distinction between FO and HO theories applies across the board to mental states of every type, as Myrto Mylopoulos (2015) shows for the case of agentive awareness, so too the distinction between one-factor and two-factor views applies to what we say about the consciousness of mental states of every mental type.

I'll argue here that 1F views face serious difficulties, difficulties that are built into those views, and so cannot be in any way avoided, adjusted for, or overcome. Any progress we might make in explaining consciousness and developing a serious scientific treatment will have to rely on 2F views.

And within the two types of two-factor views already considered, in addition GWT faces compelling counterexamples: There are many peripheral sensations that are conscious but lack any downstream psychological effects. And in the opposite direction, there are many thoughts and desires that have substantial downstream effects on psychological processing, even though those thoughts and desires are altogether unconscious. Because of that, HO theories should be our preferred type of 2F view. Still, it will be useful to cast the relevant issues in terms of the distinction between 1F and 2F views.

One especially salient difficulty that 1F views face is how to explain the way conscious states differ from mental states that aren't conscious. If a state's being conscious is intrinsic to that state, it's hard to see how a state of that same mental type could occur without being conscious.

This issue underlies the difficulty that arises with Dretske's (1993) FO claim that a state is conscious if being in it makes one aware of something. But unconscious mental states also make one aware of things, just not consciously aware. The priming effects and forced choice that subliminal perceptions often enable could not be explained without acknowledging that these unconscious perceptions make one unconsciously aware of things.

Dretske (2006) himself came to see this. But Dretske's proposed remedy for the problem illustrates vividly the difficulty in avoiding this shortcoming of 1F views without endorsing some psychological mechanism that is in 2F. And the correction Dretske does propose is plainly 2F; it argues that perceiving is conscious only if one can cite it as a justifying reason for doing something. But one cannot verbally cite something without being aware of it. This response is not only 2F, but it is in effect HO in requiring a type of HOA.

Nagel explains consciousness by appeal to cognate terms such as "what it's like," "subjectivity," "point of view" and "perspective." And on his view we cannot break out of that closed family without losing what is central to consciousness. So we cannot then explain how conscious states differ from unconscious mental states.

And all this points to a deeper difficulty. If being conscious is an intrinsic property of conscious states, confining us to Nagel's small circle of interdefined terms, not only are we unable to explain consciousness in any useful way; we can't even describe in an informative way what it is for a state to be conscious (Rosenthal 1983).

This is reminiscent of the closed curve of terms W. V. Quine (1951) sees as figuring in any attempt to explain what it is for a statement to be analytic. Just as the inability to break out of Quine's closed curve shows that any such explanation of analyticity is uninformative, so too with the inability to break out of the closed curve that Nagel insists must figure in any adequate description of consciousness.

Ned Block's conception of phenomenal consciousness is also a 1F view, denying that it can be explained in terms of anything extrinsic to a phenomenally conscious state itself. And to his credit, Block in effect explicitly acknowledges the difficulty just noted. We can, he concedes, say little if anything about what conscious qualitative character is beyond Louis Armstrong's famous quip about jazz: "If you gotta ask, you ain't never gonna get to know" (1978, p. 281). And more recently: "The best you can do is use words to point to a phenomenon that the reader has to experience from the first-person point of view" (2015, p. 47). Pointing in this context is plainly metaphorical, and so of no relevant help. How does one point to another person's experiences?

This observation generalizes. Those who claim that the mental property of being conscious is intrinsic to one or another type of mental state, or perhaps to mental states of any type, are committed to maintaining that all such mental states are conscious. And this contention precludes any informative way to describe how states of the relevant mental type that are conscious differ from those that are not.

This will encourage many 1F theorists to maintain that if a state is not conscious, it cannot genuinely be a mental state. And if that seems too stark, one might construct special conditions for unconscious mental states, such as perceptions or thoughts, as Block (2017) does for unconscious perception. But the conditions Block offers for an unconscious state to qualify as a perception do not match the conditions that define conscious perceptions. So the states he regards as perceptions in the unconscious case are states of a different mental type from conscious perceptions.

This is likely to be so for any attempt to provide conditions for a distinctive type of mental state's occurring unconsciously if one also holds that the property of a state's being conscious is intrinsic to the states in the conscious cases. And if one can't say in any useful or informative way how conscious states differ from states that are not conscious, one also won't be able to say what it is for a state to be conscious in the first place. And then the term, "conscious," becomes no more than a kind of honorific, a way of saying something nice about mental states.

Given the lack of any informative account of what it is for a state to be conscious that's cast in distinctively psychological terms, Block offers instead a neural implementation, favoring Victor Lamme's (2006) proposal that perceptions are conscious when recurrent processing occurs in sensory cortex. But a neural implementation by itself simply won't do. Consciousness is a psychological phenomenon. So we must explain in psychological terms what it is for a state to be conscious. Neural input may come to influence and adjust our initial explanation, but we must start with something psychological.

Indeed, a psychological account will typically guide our search for a neural implementation, for example, whether to look in sensory or in prefrontal or parietal areas of cortex. Because most 1F views say almost nothing about consciousness in psychological terms, they encourage instead offering accounts that are cast solely in neural terms.

And that can distort things. It suggests looking just for a neural on–off switch for conscious states, bypassing any explanation in psychological terms of what it's like for one, that is, an explanation of how exactly conscious and unconscious mental states differ. 2F views invite a more fine-grained neural implementation, which captures not only the difference between conscious and unconscious states but also, crucially, the very many different ways in which states subjectively occur in consciousness. Whatever the correct neural implementation ultimately turns out to be, it cannot be some simple on–off switch.

The focus on finding a neural correlate of consciousness is due largely to the temptation, amplified by 1F theories, of thinking that consciousness is a mysterious matter that resists any informative description, as with the foregoing remarks quoted from Block. Since consciousness is a distinctively psychological phenomenon, one would expect, and indeed hope for, an informative description cast in psychological terms. But if one deems that impossible, there is perhaps nothing informative left to say except what the neural correlates of consciousness are. (For a thoughtful review of neurobiological correlates and their theoretical status, see Mylopoulos 2022.)

The natural and obvious move here is to see whether we can dispel whatever sense there is that consciousness is indeed a mysterious phenomenon that resists informative explanation. And that is indeed that strategy of 2F views, which seeks to explain what it is for a state to be conscious by appeal to informative factors extrinsic to the conscious state itself.

Hakwan Lau (2022) has advanced a novel and ambitious 2F view that has deservedly captured widespread attention. A visual state is conscious if there is a neural state in the prefrontal cortex (PFC) that neurally points to the relevant state in the visual cortex. Absent such neural pointing by some state in the PFC, the state in the visual cortex would occur, but not consciously. This is similar with other types of perceptual state.

There are two difficulties with this view. One is that the pointing state determines only whether the visual state is conscious, not what visual properties it is conscious in respect of. It is the visual state itself, according to Lau, which determines in what way the state is conscious, for example, in respect of what color and shape. But it is arguably theoretically awkward at best to divide things up in this way, so that whether a state is conscious is determined by one neural factor and how it is conscious is determined by another. It would be more natural to explain whether a state is conscious and how it is conscious by appeal to the same considerations.

And there is another worry for this aspect of Lau's account. A perceptual state is not always conscious in respect of exactly the visible input that results from a particular stimulus. Conceptual factors may influence how a perceptual state is conscious, as when a gray banana shape is consciously perceived as yellow. In that case, visual color presumably registers the gray color, but the conscious experience is as of yellow. If the state in the visual cortex did determine how the resulting state is conscious, that would not happen. This concern arises for all cases in which subjective awareness does not accurately reflect the mental properties of the FO state, a circumstance that will be discussed independently in section 3.

Lau describes this account as a type of HO theory, since the second factor occurs in PFC and operates on the FO visual state. But on a standard HO theory, the HO represents one as being in a FO state of a particular type, and that explains both why the FO is conscious and also why it is conscious in respect of the properties that subjectively appear in consciousness. It's not obvious what theoretical advantage Lau's joint-determination account has, on which those two matters are explained independently.

Lau might urge that his view is simpler and more straightforward. But it's arguable that this is so only from the point of view of the neural underpinnings. And, as already argued, a theory of consciousness must stress and rely on those aspects of the situation that are distinctively psychological, since consciousness is itself a distinctively psychological phenomenon.

And there is a second difficulty. On Lau's theory, the HO state in the PFC operates simply to point neurally at a FO state in, for example, the visual cortex. On a standard HO theory, the HO state would represent the occurrence of that FO state, describing it in psychological terms. We know that states in the PFC have rich representational properties, often conceptual properties. But at present we are very much in the dark about how the PFC does that. There is no simple neurological map, as there is for the sensory cortical areas, which describes how neural activity at a particular location represents properties of a particular type.

So it is open that the states Lau posits state in the PFC as pointing to particular FO states actually represent one as being in the relevant state. It is hard to see how one could settle, empirically or theoretically, whether that is so. And absent some way of doing so, it's not clear that Lau's pointing model provides us with a clear alternative to more standard HO theories.

All told, the foregoing considerations point to 2F views. We must explain consciousness by appeal to phenomena that are psychological, but not themselves conscious, since explaining how one state is conscious by appeal to other states that are themselves conscious would be circular and uninformative.

And that is exactly the strategy of 2F views. GWT appeals to downstream psychological processing, which itself need not be conscious. And

HO theories rely on a HOA: A state is conscious if one is aware of it in a suitable way. And again, that HOA need not be conscious. The HOA's being conscious would require a yet third-order awareness, which itself is rare.

Indeed, we are seldom subjectively aware of any such HOAs. We know about them, from the reasoning provided by HO theories. We rarely have first-person access to them; instead, they are theoretical posits that do a good explanatory job. On 2F views a state is conscious if that second factor is present, not otherwise. That second factor is what a state's being conscious consists in.

But HO theories and GWT diverge in terms of how credible they are. If one is wholly unaware of a mental state one is in, that state is not conscious. This is basic to our commonsense conception of consciousness, and indeed also central to experimental methodology in consciousness research. And that points to a HO theory, on which a state is conscious only if one is in some suitable way aware of it.

By contrast, it is unclear why one would expect that, as a general matter, a state's being conscious consists in its content being available for downstream processing. Indeed, the counterexamples to GWT noted earlier suggest otherwise. Many conscious states do, of course, influence downstream psychological processing; they are typically those conscious states that involve some measure of attention. But consciousness and attention also occur altogether independently of one another (Norman et al. 2013). And it is unclear what in the nature of consciousness, independently of attention, would in general result in a conscious state's having an influence on downstream processing.

I'll come back to this issue shortly.

3   THEORY AND APPEARANCE

A 2F view readily lends itself to scientific treatment, since one can empirically investigate and theorize about each factor independently of the other. For GWT, one can study the potential that various states have for their contents to result in downstream processing, and isolate the considerations, such as signal strength, that affect that potential. For HO theories, one can investigate and theorize about the kind of HOA that figures in a state's being conscious and determine what considerations give rise to that HOA and how.

By contrast, the 1F view that consciousness is an intrinsic property is far less conducive to theory and to empirical investigation. In part this is because there are fewer moving parts. If a property is intrinsic, it's less available for the manipulations that figure in experimentation and theorizing. If one thinks of the mental circumstances in virtue of which the state is conscious as intrinsic to the state, there is only one mental factor to investigate. Indeed, construing a property as intrinsic is akin to seeing it as an essence, and so as something that is simply given, and so resistant to informative explanation.

But there is a deeper reason that 1F views are recalcitrant to theorizing and empirical investigation. If consciousness is intrinsic, it's the last word about the nature of every conscious state. Consciousness then overrides any other information we could possibly have about the mental properties of the state in question. So as already noted, seeing consciousness as intrinsic detaches it from everything we might investigate independently of first-person access, so that there is then nothing objective to say about the nature of consciousness or how it comes to be.

Once we construe being conscious as intrinsic to those mental states that are conscious, what we can know about conscious states in psychological terms is limited, as the foregoing quotations from Block in effect put it, to saying that this is what it's like, where 'this' refers to something accessible only to the subject.

It is this closed-off, consciousness-first picture that by itself gives rise to a sense of mystery about consciousness, which can express itself in the claim that there is an explanatory gap (Levine 2001) or a hard problem (Chalmers 2003). A 2F view, by contrast, opens up rich ties that being conscious has with other mental phenomena, and the availability of such ties dispels those apparent mysteries. More about the hard problem and explanatory gap shortly.

Because on 1F views we know about the nature of conscious states exclusively from what consciousness itself tells us, 1F views are strongly anti-scientific and anti-theoretical. Thus, Nagel (1974) insists that if our description of conscious mentality included anything objective, it would, solely because of including an objective factor, fail to do justice to the nature of consciousness as pure subjectivity.

Other 1F advocates rarely explicitly acknowledge this anti-theoretical aspect of their view. In part that is because they compensate for the lack of any theoretical account in distinctively psychological terms by pursuing direct correlations of conscious occurrences with neurological occurrences, which in effect distracts attention from the lack of a genu-inely theoretical account.

But one can see that anti-theoretical attitude inherent in 1F views by the way their advocates simply ignore the theoretical aspect of HO views, and instead measuring all views solely against the subjective appearances. An example of this occurs in connection with the issue about whether a HOA might misrepresent the state it makes one subjectively aware of.

1F advocates note that on a HO theory a HOA might misrepresent a conscious state, and they regard this as decisive against HO theories (e.g. Byrne 1997; Levine 2001; Neander 1998). Such misrepresentation does actually occur. For example, in change blindness one is often remains con-sciously aware of the stimulus or object that changed in the way it was prior to the change (Grimes 1996; Fallon 2020–22). But since the change is often unconsciously perceived (Fernandez-Duque and Thornton 2000; Thornton and Fernandez-Duque 2001; Laloyaux et al. 2006), it is plain that the visual cortex often registers the changed properties after the change has taken place. So conscious awareness misrepresents the post-change visual state.

Subjective misrepresentation is never evident simply from the subjec-tive appearances themselves, since the subjective appearances are them-selves misrepresenting. Though the subjective appearances on which 1F views rest cannot reach beyond themselves, HO theories do reach beyond those appearances. It is that feature of HO theories that 1F views find objectionable. HO theories actually theorize about the subjective appear-ances themselves.

One can describe such subjective misrepresentation in a way that makes it seem problematic. Suppose a change-blindness subject misses a change of something from green to red, as in the parrot display in Grimes (1996), and continues to see a green object. HO theories could describe that as having a conscious sensation of green. But if the post-change input causes a red state in the visual cortex, there is no green sensation to begin with. How can one have a conscious green sensation but no green sensation?

Still, as Daniel Shargel (2016) nicely stresses, describing the situation as being problematic in that way is not only highly misleading, but also alto-gether optional. The accurate description is that this change-blindness subject has an unconscious red sensation, but it subjectively appears to the subject as though there is a green sensation (see also Berger 2014). The subjective appearances in such a case simply do not accurately reflect the sensory state of the individual.

And there is nothing obviously problematic about that. Indeed, we can sometimes discriminate in a more fine-grained way than subjective awareness is able to capture, even though all the relevant perceptual states are conscious. This is evidently what happens in an experiment described by Diana Raffman (2011), in which participants subjectively judge that pairs of color patches are identical in shade, even though they match the two samples in a way that accurately reveals the difference between them. In this case subjective awareness simply does not reflect our sensory states with full accuracy, and this doubtless also happens very often in everyday perceptual experience.

1F advocates also note we're rarely aware subjectively of any HOAs, ignoring that HOAs are theoretical posits, and indeed are posited as rarely themselves being conscious. We can again see how 1F advocates reject theoretical thinking about the nature of consciousness. In confining themselves to the subjective appearances, 1F views seriously distort the way we actually think about consciousness and psychological functioning.

This resistance to theorizing also affects the way many 1F advocates think about what utility conscious states have that is due exclusively to the property of those states being conscious. Much of a conscious state's utility must, of course, be due to its content properties, which drive most of its causal connections with behavior and other states, since differences in content properties result in different results (Rosenthal 2008). But confining attention to the subjective appearances encourages running together the content properties with consciousness, which leads to a sense that all the utility is due instead to consciousness.

And though a GWT is not itself 1F, this kind of thinking about the utility of a state's being conscious may well encourage a GWT, on which a state's being conscious is tied to its downstream psychological results. This connection is especially vivid in the work of Danel C. Dennett on consciousness (e.g., 1991), which itself is arguably 1F in rejecting any HO notion of seeming (Rosenthal 2018), and accordingly ends up explaining consciousness by appeal to its behavioral effects.

Yet another way to see the anti-theoretical aspect of 1F approaches to consciousness pertains to the so-called explanatory gap and hard problem mentioned earlier. It's said that there is a hard problem about consciousness because of the apparent difficulty in explaining how neural processes could give rise to, or actually constitute, conscious experiences. And that difficulty can be described as an explanatory gap that blocks explaining consciousness by appeal to neural processes.

The apparent difficulty in giving such an explanation is usually described by appeal to the intuitive difficulty in seeing how neural processes could result in consciousness. It is said that we can imagine or conceive of those neural processes without consciousness, and that nothing about them seems intuitively to result in consciousness. It doesn't seem to make intuitive sense that they would.

But does it make intuitive, pretheoretic sense that a liquid would result from bonding together two gases, as with water? Or that most of the spatial extension of every solid object should be constituted by nothing solid, but at most wave-mechanical phenomena We don't assess scientific explanations by whether they make intuitive, pretheoretic sense, but whether they're embedded in a body of theory that generates solid predictions and fits comfortably with other bodies of theory.

Those who see a hard problem or an explanatory gap will argue that consciousness is a special case, in part because they maintain that we can imagine or conceive of whatever neural processes figure in a proposed explanation without any conscious experiences. But we can also conceive of hydrogen and oxygen being bonded in the right way without there being any water. What we can't conceive is that could happen compatibly with the relevant science that we know. And once we have a good predictive theory of consciousness and apply that standard, we won't be able in that way to conceive of the relevant neural processes without conscious experiences.

The appearance of a hard problem or an explanatory gap is fed in part by the odd view that an explanation of conscious mental phenomena should proceed directly from neural processing. But, as already noted, this demand is reasonable only if one has already ruled out anything informative to say about consciousness in distinctively psychological terms. So one shouldn't appeal to the hard problem or explanatory gap in urging that the only explanation of consciousness must be directly in terms of neurological functioning. Indeed, since consciousness is plainly a psychological phenomenon, one would expect any explanation to be cast in the first instance in psychological terms, only moving to the neurological level when the psychological factors are all in place. Skipping over those intervening psychological factors is part of what makes it seem that a neurological explanation of consciousness doesn't make intuitive sense.

And there is more. Explaining consciousness by appeal to neurological occurrences will not make intuitive sense if we start with the subjective appearances of consciousness and attempt to reason to the appropriate neurological factors. It will inevitably seem that we are leaving out what is important to consciousness.

But that is always the way it is with scientific explanations. Consider explaining water by appeal to H_2O. That will seem intuitively credible when we work up from H_2O, together with chemical theory, to explain the macroscopic nature of water. Proceeding in that direction, the explanation makes good intuitive sense. But it would be hopeless to start instead with the macroscopic nature of water and attempt to work back down to the H_2O.

More generally, post-Galilean scientific explanations can engender an unintuitive feel when one starts with specific macroscopic explananda and attempts to reason to the underlying reality posited by the relevant scientific theory. We readily dispel that unintuitive feel if instead we reason from the underlying reality, together with the relevant body of theory, to the macroscopic natures they explain.

The unintuitive feel engendered by taking the macroscopic natures as basic is due in part to leaving out the relevant body of theorizing, but not entirely. It is also because the macroscopic nature of things often does not in itself contain pointers to the relevant underlying reality. Indeed, in a pre-Galilean age, taking macroscopic natures as basic encouraged an Aristotelian approach to science, on which the main goal is to describe and taxonomize those macroscopic appearances.

These considerations readily carry over to explaining consciousness. We must not think of working from the subjective nature of consciousness down to factors that explain that nature, whether psychological or neurological. We must anticipate having psychological and neurological explanations that are good enough to enable us reason from them to specific subjective appearances, as we do with H_2O and water. And the specifically neurological explanations will only be good enough for that when all the relevant psychological factors are also in place. Insofar as claims that there is a hard problem or explanatory gap rest on taking the macroscopic appearances as dictating the possible avenues of theoretical explanation, those mysterian ideas are in effect a throwback to an Aristotelian science.

Indeed, we can readily explain why it seems at all tempting to hold, with 1F views, that consciousness is intrinsic to perceptual states by appeal to a well-entrenched, but deeply faulty misconstrual of the implications of post-Galilean science. Many have taken the Galilean dictum that the book of "the universe … is written in the language of mathematics" (Galilei 1623, passage tr. Popkin 1966, p. 65) to exclude properties such as colors from the natural order. The standard reaction, typically unstated, is to relocate such properties to the mind (Rosenthal 1999).

But colors resist mathematical treatment only as they appear in conscious perception; physical colors as they are on their own, independent of being consciously perceived, pose no difficulty for a mathematical description. And if the color properties we relocate to the mind are construed as they appear in conscious perception, the relocated mental properties will again themselves have consciousness built in. It is this sleight of hand that results in the appeal of the 1F insistence that mental color qualities are intrinsically conscious (Rosenthal 2024, forthcoming, secs. III and VII). And these considerations suggest that rejecting 1F views will be essential to completing the Galileo scientific revolution.

The foregoing can be usefully supplemented by Josh Weisberg's (2024) rich discussion that carefully examines the reasons on offer for thinking that there is a hard problem, in particular, by David Chalmers (2003). Weisberg argues forcefully that those aspects of the appearances of consciousness that might seem problematic result simply from the sensory information accessed in consciousness being compressed and being accessed in an automatic way. But there is nothing problematic about such compression and automatic access. So this explanation dispels the sense that the appearances of consciousness generate a hard problem.

The foregoing considerations call for a careful examination of the popular appeal to intuitions that figures so prominently in arguments for a hard problem or explanatory gap. People don't all report having the same intuitions. More important, if one knows which intuitions somebody avows in this area, one can reliably predict what relevant theoretical positions that person will hold, and conversely.

That strikingly strong correlation suggests that intuitions are simply inviting one-liners that encapsulate a particular theoretical outlook. How could we otherwise explain the overwhelming reliability of the correlation? One could suggest that people start with intuitions and then generate a theory that fits. But where did the intuitions come from if there was no prior tendency to hold the relevant sort of theoretical position?

And if intuitions are encapsulated versions of theoretical positions, we should simply set them aside altogether and concentrate on evaluating the relevant theoretical positions. Since the intuitions themselves actually conceal the relevant theoretical thinking, focusing on them distracts from the theoretical assessment that is all important. In concentrating on what is intuitive, 1F theorists again reveal their anti-theoretical stance.

Scientific theories rarely depart totally from the macroscopic appearances. We do commonly think of things as consisting of much smaller parts; the chemical theory of water simply gives a somewhat surprising account of the relevant parts and how they're combined.

2F theories of consciousness follow suit. On a HO theory, for example, a state is conscious in virtue of one's being aware of being in that state. That fits with, and indeed is equivalent to, the commonsense idea noted earlier that no state is conscious if one is altogether unaware of being in it. What's surprising is that the HOA posited by HO theories is not itself a conscious state. It is an unconscious awareness, on a part with the subliminal awareness we have of stimulus when we perceive them unconsciously. So, a HO explanation matches other scientific explanations pretty well, relying partly on commonsense ideas about the nature of what is being explained, but supplementing and adjusting its theoretical posits for explanatory purposes.

This arguably holds also for the specific appeal to higher-order thoughts that I have advanced elsewhere (Rosenthal 2005, esp. chs 1, 2, 4, and 10–12). The theory posits a HOA in keeping with the commonsense idea that no state of which one is wholly unaware is conscious. But the theory then invokes various theoretical considerations to argue that this HOA is a conceptual state, on a par with the unconscious thoughts we often have.

And the other major type of 2F theory, GWT, also proceeds in just this way. GWT arguably starts from the commonsense idea that many, perhaps most, typical cases of conscious states have some significant utility, and posits that the downstream psychological processing that implements such functionality is what it is for those states to be conscious. Both types of 2F theory operate in the manner of standard scientific theories.

1F views are currently widely dominant in current discussion of consciousness. Given all these downsides, why are they so popular?

1F advocates contend that their picture of consciousness is well-entrenched in common sense, and so does justice to the way we think about consciousness in ordinary, everyday terms. Thus, their ubiquitous reliance on allegedly pretheoretic intuition.

1F views do start from a correct observation about our commonsense views, but then wildly overshoot. Consciousness is mental appearance. It is the way our mental lives subjectively appear to us. This is basic to any serious way of thinking about consciousness, whether in common sense or in a scientific theory.

But 1F views extrapolate from that commonsense observation to the extravagant claim that the mental appearances also exhaust the mental reality of consciousness. If this were so, there could be for consciousness no coherent distinction between appearance and reality when it comes to consciousness. And that is what Nagel famously maintains: "The idea of moving from appearance to reality seems to make no sense" for conscious experiences (p. 444).

If the mental appearances of consciousness did exhaust its mental reality, all mental reality would lie behind a wall of subjectivity, separated from anything objective. But there are, in addition to the mental appearances, objective mental realities that underlie and explain the appearances. There are the mental occurrences that subjectively appear to us, states with conceptual content or qualitative character, both of which we can theorize about apart from consciousness (Rosenthal 1986, 2010). Those states are one part of the mental reality that underlies and helps explain the mental appearances.

And there are also the mental occurrences that actually implement the subjective appearances themselves. And these are also an aspect of the objective mental reality of conscious states (Berger 2017; Rosenthal 2022, forthcoming). On GWT, that additional mental reality is the downstream availability of content. On HO theories it is the HOA of the states themselves, which represent the states specifically in respect of those mental properties that figure in our subjective awareness.

1F advocates will object to all this, and claim that distinguishing appearance from reality has no place with consciousness. But all cases of appearing are implemented by something objective. One's appearance in a mirror is not something outside the natural order; it is a matter of how light reflects off particular surfaces. The only reason to think that the appearances of consciousness are not themselves implemented by objective occurrences, which invite straightforward explanation in standard scientific terms, is that it doesn't subjectively seem that the appearances of consciousness are thus implemented.

But it wouldn't seem so subjectively: Objective implementation of all appearances is never evident from the appearing itself. We can't tell from the appearances themselves whether there is any such objective implementation. To determine what the objective implementation is in a case of appearance of any sort whatsoever we must always appeal to factors that lie outside the appearances themselves.

The idea that we could determine how a type of appearance is implemented from considerations internal to the appearances is of a piece with the idea noted above that we can assess the credibility of a scientific explanation of consciousness by seeing whether we can reason from how things seem subjectively to the factors appealed to in the explanation. It is the mistake of thinking that how thing appear must govern any account of those appearances.

That is what leads to 1F advocates rejecting any appearance–reality distinction for consciousness, and to the illusion (Mandik 2016; Weisberg 2024) that there is a hard problem or an explanatory gap. Rejecting an appearance-reality distinction for consciousness is groundless, and with it the adoption of a 1F view. We must accept and work with a 2F view.

REFERENCES

Baars, Bernard J. 1997. *In the Theater of Consciousness: The Workspace of the Mind*, Oxford: Oxford University Press.

Berger, Jacob. 2014. Consciousness is Not a Property of States: A Reply to Wilberg,, *Philosophical Psychology*, 27, 6: 829-842.

Berger, Jacob. 2017. How Things Seem to Higher-Order Thought Theorists, *Dialogue: Canadian Philosophical Review*, 56, 3: 503-526.

Block, Ned. 1978. Troubles with Functionalism, in *Minnesota Studies in the Philosophy of Science*, IX, ed. C. Wade Savage, Minneapolis: University of Minnesota Press, pp. 261-325.

Block, Ned. 2015. The Puzzle of Perceptual Precision, *In* Thomas Metzinger and Jennifer M. Windt (Eds), *Open MIND*, Frankfurt am Main: MIND Group, doi: https://doi.org/10.15502/9783958570726.

Block, Ned. 2017. Unconscious Perception within Conscious Perception, 7-9 of 'Does Unconscious Perception Really Exist? Continuing the ASSC20 Debate, by Megan A. K. Peters, Robert W. Kentridge, Ian Phillips, and Ned Block, *Neuroscience of Consciousness*, 1, nix015, https://doi.org/10.1093/nc/nix015.

Byrne, Alex. 1997. Some Like It Hot: Consciousness and Higher-Order Thoughts, *Philosophical Studies*, 86, 2: 103-129.

Chalmers, David J. 2003. Consciousness and Its Place in Nature, *In* Stephen P. Stich and Ted A. Warfield (Eds), *The Blackwell Guide to Philosophy of Mind*, Massachusetts: Blackwell Publishing Ltd., pp. 102-142.

Dehaene, Stanislas, and Lionel Naccache. 2001. Towards a Cognitive Neuroscience of Consciousness: Basic Evidence and a Workspace Framework, *Cognition* 79, 1-2: 1-37.

Dennett, Daniel C. 1991. *Consciousness Explained*. Boston: Little, Brown and Company.

Dretske, Fred. 1993. Conscious Experience, *Mind*, 102, 406: 263–283.

Dretske, Fred. 2006. Perception without Awareness, *In* Tamar Szabó Gendler and John Hawthorne (Eds), *Perceptual Experience*, Oxford: Clarendon Press, pp. 147-180.

Fallon, Francis, Alan L. F. Lee, Brian Odegaard, Andrew Haun, and David Rosenthal. 2020-2022. Replication and Extension of Crucial John Grimes Experiment: Change Detection during Saccades, *Templeton World Charity Foundation Project* 0455.

Fernandez-Duque, Diego, and Ian M. Thornton. 2000. Change Detection without Awareness: Do Explicit Reports Underestimate the Representation of Change in the Visual System?, *Vis. Cog.*, 7, 1-2-3: 324–344.

Galilei, Galileo. 1623. *Il Saggiatore*, Rome.

Grimes, John. 1996. On the Failure to Detect Changes in Scenes across Saccades, In Kathleen Akins (Ed.), *Perception*, New York: Oxford University Press, pp. 89–110.

Lamme, Victor A. 2006. Towards a True Neural Stance on Consciousness, *Trends in Cognitive Sciences*, 10, 11: 494-501.

Laloyaux, Cédric, Arnaud Destrebecqz, and Axel Cleeremans. 2006. Implicit Change Identification: A Replication of Fernandez-Duque and Thornton (2003), *Journal of Experimental Psychology: Human Perception and Performance*, 32, 6: 1366-1379.

Lau, Hakwan. 2022. *In Consciousness We Trust: The Cognitive Neuroscience of Subjective Experience*, Oxford: Oxford University Press (open access: chrome-extension://efaidnbmnnnibpcajpcglclefindmkaj/http://fdslive.oup.com/www.oup.com/academic/pdf/openaccess/9780198856771.pdf).

Levine, Joseph. 2001. *Purple Haze: The Puzzle of Consciousness*, New York: Oxford University Press.

Mandik, Pete. 2016. Meta-Illusionism and Qualia Quietism, *J. of Consciousness Studies*, 23, 11-12: 140-148.

Michel, Matthias, and Adrien Doerig. 2022. A New Empirical Challenge for Local Theories of Consciousness, *Mind & Language*, 37, 5: 840-855.

Mylopoulos, Myrto. 2015. Consciousness, Action, and Pathologies of Agency, *In* Rocco J. Gennaro (Ed.) *Disturbed Consciousness: New Essays on Psychopathology and Theories of Consciousness*, Cambridge, Massachusetts, and London: MIT Press, pp. 75-102.

Mylopoulos, Myrto. 2022. Neurobiological Theories of Consciousness, *In* Benjamin D. Young and Carolyn Dicey Jennings (Eds), *Mind, Cognition, and Neuroscience: A Philosophical Introduction*. New York and London: Routledge, pp. 280-293.

Naccache, Lionel. 2018. Why and How Access Consciousness Can Account for Phenomenal Consciousness, *Philosophical Transactions of the Royal Society B: Biological Sciences*, 373, 1755: Article 20170357.

Nagel, Thomas. 1974. What Is It Like to Be a Bat?, *The Philosophical Review*, 83, 4: 435-450.

Neander, Karen. 1998. The Division of Phenomenal Labor: A Problem for Representational Theories of Consciousness, *Philosophical Perspectives*, 12: 411-434.

Norman, Liam J., Charles A. Heywood, and Robert W. Kentridge. 2013. Object-Based Attention without Awareness, *Psychological Science*, 24, 6: 836-843.

Popkin, Richard H. 1966. *The Philosophy of the Sixteenth and Seventeenth Centuries*, New York: Free Press.

Quine, W. V. 1951. Two Dogmas of Empiricism, *The Philosophical Review*, 60, 1: 20-43.

Raffman, Diana. 2011. Vagueness and Observationality, *In* Giuseppina Ronzitti (Ed.), *Vagueness: A Guide*, Dordrecht: Springer, pp. 107-122.

Rosenthal, David. 1983. Reductionism and Knowledge, *in* Leigh S. Cauman, Isaac Levi, Charles Parsons, and Robert Schwartz (Eds), *How Many Questions? Essays in Honor of Sidney Morgenbesser*, Indianapolis: Hackett Publishing Co., pp. 276-300.

Rosenthal, David. 1986. Intentionality," *Midwest Studies in Philosophy* X: 151-184.

Rosenthal, David. 1999. Sensory Quality and the Relocation Story, *Philosophical Topics* 26, 1/2: 321-350.

Rosenthal, David. 2005. *Consciousness and Mind*, Oxford: Clarendon Press.

Rosenthal, David. 2008. Consciousness and Its Function, *Neuropsychologia*, 46, 3: 829-840.

Rosenthal, David. 2010. How to Think about Mental Qualities, *Philosophical Issues*, 20: 368-393.

Rosenthal, David. 2018. Seeming to Seem, *In* Bryce Huebner (ed.), *The Philosophy of Daniel Dennett*, New York: Oxford University Press, pp. 133-164.

Rosenthal, David. 2022. Mental Appearance and Mental Reality, *In* Josh Weisberg (Ed.), *Qualitative Consciousness: Themes from the Philosophy of David Rosenthal*, ed, Cambridge University Press, pp. 243-271.

Rosenthal, David. 2024. Methodological Considerations for the Study of Mental Qualities, *In* Juraj Hvorecký, Tomáš Marvan, and Michal Polák (Eds)., *Conscious and Unconscious Mentality: Examining Their Nature, Similarities and Differences*, London and New York: Routledge.

Shargel, Daniel. 2016. The Insignificance of Empty Higher-Order Thoughts, *Journal of Cognition and Neuroethics*, 4, 1: 113–127.

Thornton, Ian M., and Diego Fernandez-Duque. 2001. An Implicit Measure of Undetected Change, *Spatial Vision*, 14, 1: 21-44.

Weisberg, Josh. 2024. *Explanatory Optimism about the Hard Problem of Consciousness*, New York and London: Routledge.

CHAPTER 7

Theoretical Virtues of Cognitive Extension

Marcin Miłkowski and Juraj Hvorecký

1   INTRODUCTION

The goal of this chapter is to argue that the extended mind approach can be distinguished from its alternatives in terms of not only metaphysics, but also epistemology. In other words, we claim that this view differs in its theoretical virtues compared to other similar approaches to cognition.

To accomplish this, we will proceed as follows: In Sect. 2, we will introduce the objection against the extended mind view on cognition that

Juraj Hvorecký: Work on this chapter was supported by the Czech Science Foundation (GAČR), project n. 20–14445S ('Dual Models of Phenomenal Consciousness') realized at the Institute of Philosophy, Czech Academy of Sciences.

M. Miłkowski (✉)
Institute of Philosophy and Sociology, Polish Academy of Sciences, Warsaw, Poland
e-mail: marcin.milkowski@ifispan.edu.pl

J. Hvorecký
Institute of Philosophy, Czech Academy of Sciences, Prague, Czech Republic
e-mail: hvorecky@flu.cas.cz

P. Alexandre e Castro (ed.), *Challenges of the Technological Mind*, New Directions in Philosophy and Cognitive Science, https://doi.org/10.1007/978-3-031-55333-2_7

states that it does not differ from its competitors in terms of the empirical content of the explanations involved. We will use the example of the audience effect as framed in terms of extended emotional processing to illustrate this point. Then, we will argue that there are further criteria of theory choice, traditionally understood in terms of theoretical virtues, that are implied by the extended view and its alternatives. Specifically, we will argue that previous discussions concerning the advantages and disadvantages of the approach lead to the observation that it may have problematic consequences for the generality and simplicity as well as unificatory properties of cognitive theories. In this regard, the extended view, particularly in its more recent versions, has specific epistemic implications for the development of cognitive theories. At the same time, we will note that the theoretical virtues of general approaches to cognition (or theoretical frameworks) should not be confused with those of particular explanations of phenomena. In conclusion, we will state that the extended view is epistemically distinguishable from its competitors. Therefore, the objection can be rebutted.

However, our conclusion is not particularly optimistic for the view, as the difference does not seem to be in its favor due to considerations of generality, parsimony, and unification. While it is certainly possible for defenders of the extended cognition to develop their theoretical framework to warrant a more systematic approach to developing explanations that remain appropriately general, parsimonious, and unificatory, the way this generality, parsimony and unification could be achieved will most likely be quite different from traditional approaches that rely on invariant structures in individuals or cognitive systems to warrant these desiderata.

2   THE ISSUE OF EMPIRICAL EQUIVALENCE

In this section, we will address the issue of the empirical equivalence of the extended mind approach with similar, but metaphysically distinct frameworks in contemporary cognitive science. Essentially, if the extended mind view is indistinguishable from alternative views in terms of empirical evidence, then the utility of engaging in the metaphysics of the extended mind could be called into question and seen as a futile exercise (Müller 2018). Additionally, some have argued that there is no empirical difference between the situated cognition approach and the extended mind approach (Barker 2010; Sprevak 2010). Barker states that differences between the two approaches are not empirically testable and that they cannot be distinguished based on theoretical virtues, in particular simplicity and unification. In short, they are equivalent in terms of their empirical accuracy, testability, and other theoretical virtues. Sprevak argues that by looking at the cognitive science practice, one cannot actually settle the debate between the two approaches, as cognitive science is not sensitive to the difference between causal and constitutive relevance. If these arguments hold, then it becomes unclear what, if any, relevance the extended mind view holds for empirical research. This also undermines the original claims of Clark and Chalmers, who stated that their view had "obvious consequences for philosophical views of the mind and for the methodology of research in cognitive science" (Clark and Chalmers 1998, p. 18).

Before we can assess this claim, it is important to clarify how we understand the terms "situated cognition," "extended cognition"[1] and "distributed cognition." These terms all refer to approaches that fall under the umbrella of 4E cognition (embodied, embedded, extended, and enactive), or wide cognition (Miłkowski et al. 2018). They are also sometimes linked to the theory of niche construction, which posits that an organism's choices, activities, and metabolic processes shape or modify the environment (Laland et al. 2000; for a more detailed examination in the cognitive and emotional domain, see Krueger 2014). The situated approach to cognition, also known as "embedded cognition," holds that cognition should be understood in terms of the (typically time-sensitive) interaction between the agent and its immediate surroundings. According to it, the extrabodily context constrains and enables cognition. The extended mind idea suggests that cognitive processes are not necessarily limited to the brain and can incorporate external resources such as tools, language, and external systems in the environment. Resulting cognitive processes literally incorporate the external elements as their integral parts. Finally, the distributed cognition approach does not focus on cognition as a property of an individual organism or agent. Rather, it describes larger cognitive systems, which may encompass multiple individual agents and artifacts. Within the distributed framework, cognitive states often result from interactions among several agents and are shared among them.

At first glance, these approaches may appear to differ in their claims. However, upon closer examination, it may be that there is no objective way to distinguish their explanatory achievements. For example, let's

---

[1] In this chapter, we use the terms "extended mind" and "extended cognition" interchangeably.

consider an agent A, who employs a tool T_1 in their cognitive process. According to situated cognition, T_1 is considered part of the environment, whereas, in extended cognition, it is regarded as part of A's mind. Under the framework of distributed cognition, T_1 is seen as part of a larger cognitive system that encompasses A and T_1. Nonetheless, the causal impact of T_1 remains the same across all three cases. Within all three frameworks, the attainment of a specific goal would not have been possible without the interaction between A and T_1. In other words, the causal structure responsible for the cognitive process remains the same in all three cases, with the only difference lying in the labels attributed to A and T_1. At best, these labels could be seen as interpretive glosses, but their theoretical significance is questionable as their explanations proceed in the same manner. Thus, the argument can be presented as follows:

1. Extended, situated, and distributed explanations of a cognitive phenomenon C posit the same causal structures responsible for C.
2. If an explanation E_1 posits the same causal structure for C as E_2, E_1 is explanatorily (epistemologically) equivalent to E_2.

Hence, extended, situated, and distributed explanations of C are explanatorily equivalent.

This is an impeccable *modus ponens*. While in actual cases, the specific approaches may differ in the proposed causal structures, this is only because researchers tend to make additional causal assumptions. In reality, all situated cognition explanations could be rephrased in terms of extended or distributed cognition and vice versa without any loss of empirical meaning. This implies that the entire debate over the extended mind or distributed cognition is purely metaphysical. This may put supporters of wide cognition in a difficult position, as it suggests that the extended mind thesis is indeed purely a matter of terminology (Müller 2018).

However, there is a problem with the argument presented above. The issue is that causal structures are always posited to explain a particular phenomenon, and how one defines the phenomenon to be explained is always influenced by one's theoretical perspective (Craver 2009). All three approaches, if fact, imply different theoretical perspectives on cognitive phenomena. This means that their definitions of the phenomena to be explained will inevitably differ, and that thus the causal structures must be different as well. We can illustrate this with the example of the audience effect, which can be understood through the lens of extended emotional processing (Krueger 2014; Krueger and Szanto 2016).

The "audience effect" (Fernandez-Dols and Ruiz-Belda 1997) is grounded in the observation that individuals exhibit more pronounced emotional expressions in the presence of others compared to when they are alone. This effect is particularly evident in sport situations where strong emotions and crowds are both present. For example, it has been observed that upon a successful strike, bowling players tend to smile more intensely when turning toward their fellow players, rather than when hitting the pins (Kraut and Johnston 1979). This is also the case in what may be considered the pinnacle of a sporting career, such as winning at the Olympics. During award ceremonies, athletes tend to express the most joy during the handshake with an Olympic official, when social interaction is most pronounced. They are significantly less likely to express their joy when alone, without the presence of a perceptive crowd. In this regard, it would be interesting to compare the expressions of joy of athletes during regular award ceremonies to similar situations without crowds (such as the Tokyo Olympics of 2021 due to the COVID-19 situation). The audience effect, which requires external stimulation, stands in sharp contrast to internalist theories of emotions that emphasize the role of mandatory emotional expression stemming solely from within the organism (Ekman et al. 2013). Ekman and his collaborators take emotion expressions to be externally directed demonstrations of inner processes that take place within the individual, fully in line with internalist principles.

The key finding from studies on the audience effect is that the intensity of emotional reactions, as seen in facial expressions, is influenced by social interactions and reinforced by the audience to whom it is directed. However, it is not a one-way projection from the athlete to the crowd. Instead, the emotion is evoked by the presence of the audience, making the emotion expression stronger for the athlete. The emotions under discussion are a result of the coupling between individual athletes and onlookers. This does not mean that people do not experience emotional episodes when alone. The effect only shows that the intensity of emotion expression varies with the presence of involved observers. There are even suggestions that in solitary scenarios, implicit sociality in the form of imagined or hypothesized observers can enhance emotional reaction (Fridlund 1991).

The audience effect serves as a prime example of a mental phenomenon that surpasses the boundaries of the individual. All three frameworks, of embedded, extended, and distributed cognition, have the capability to account for this phenomenon, albeit from different perspectives. The embedded mind framework directs its attention toward the situated nature of the individual mind within a specific environment, such as a stadium, to comprehend the effect. On the other hand, the extended mind approach proposes that the expression of emotions is co-constructed by the cheering audience, in addition to the internal input from the athlete. On this view, members of the crowd are actively involved in the experience of joyful emotion, contributing to its sustenance. Meanwhile, the perspective of distributed cognition posits that the mental state of triumphant joy is both located and shared among all participating actors, including the athletes, the crowd, and other relevant stakeholders.

As we've seen, the theoretical posits each of the three frameworks bring in differ substantially, but these differences do not prevent each framework from explaining the audience effect phenomenon successfully. In fact, even from the brief description of their explanatory strategies, it becomes clear that the main difference remains largely metaphysical. Each approach introduces a different ontology for mental entities and offers distinct conditions for their individuation and persistence. However, Barker (2010) argued that these differences were not testable. The varying interpretations of the effect are yet another reason to express skepticism about the precise ontology of mental states, as well as the futility of the concurring frameworks. Everyone agrees that the phenomenon cannot be explained without reference to external events. Thus, the precise understanding of the nature of mental states is, at best, secondary.

Before we proceed, an important caveat is in order. Extended or embodied cognition, as frameworks, possess limited (if any) inherent explanatory power. However, this is not fundamentally different from other types of frameworks, such as computationalism (Miłkowski 2018; Wołoszyn and Hohol 2017). The reason is that, as frameworks, their primary role is not to directly explain specific phenomena. Instead, they serve as inspiration and guidance for researchers in constructing general explanatory theories or specific models for the phenomena under investigation. As we will argue below, this caveat isn't a mere terminological quibble; there are deeper nuances to explore.

In this section, we have outlined the problem of the empirical equivalence of the extended mind approach with similar, but metaphysically distinct frameworks in contemporary cognitive science. We have shown that, despite different theoretical posits, all three frameworks can successfully explain the audience effect phenomenon. However, this leaves us with a purely metaphysical debate, with no clear way to test the ontological differences between the frameworks. While some may argue that this ends the discussion, we argue that by examining theoretical virtues, there may be reasons to prefer one version of the wide approach over others, *contra* Barker. In the next section, we will examine this issue in greater detail.

3   PROBLEMS WITH GENERALITY

In this section, we argue that even though the extended mind approach and its alternatives may exhibit explanatory equivalence, there are still reasons to prefer certain versions of a particular approach over others. To make this case, we will examine the theoretical virtues of these approaches. These virtues include accuracy, breadth of scope (or generality), consistency, simplicity, and fruitfulness, as outlined by philosophers of science such as Kuhn (1977). While for space reasons, we won't be able to examine all of these virtues in depth, we will provide a brief overview of each of them to set the stage for their later use in our analysis. Subsequently, we will focus on generality and unification.

Accuracy relates to a theory's ability to adequately account for specific phenomena and its alignment with various experimental results. Consistency can be evaluated in both internal and external forms. Internal consistency assesses whether a theory's claims are non-contradictory, while external consistency concerns its compatibility with other accepted theories within the same field and beyond. The breadth of scope refers to the range of phenomena that a theory addresses. Simplicity relates not only to the number of entities introduced and the manageability of the theoretical apparatus, but also to the assumption that the theory introduces order among previously divergent phenomena. Finally, fruitfulness refers to a theory's ability to discover new phenomena and/or relationships between already known ones, as well as its potential for further productivity. This criterion is synonymous with a theory's ability to initiate or sustain a viable research program (Lakatos 1970) or tradition (Laudan 1977).

We are aware that in the course of his thinking on the strength and utility of the five criteria, Kuhn expresses skepticism about an unproblematic way to make simple comparisons between competing theories (see, for example, Okasha 2011). Furthermore, Kuhn's list is not definitive, and in the course of further discussions, longer and more structured lists have been produced (for a recent one, see Keas 2018). We do not claim that a particular list is complete, nor that there might not be disputes about how to interpret each criterion. Our argument is that there are some *epistemo-logical* criteria for theory choice that go beyond explanatory adequacy and purely metaphysical debates, and they might prove useful when employed in the evaluation of competing approaches. For us, Kuhn's original list is a starting point for a debate that has implications for comparisons between situated, extended, and distributed mind frameworks. As we aim to show, on some criteria all frameworks come out the same, but there are dimensions in theory choice where substantial differences appear, and some approaches will score low on them, which might have consequences for their broader acceptability.

In the remainder of this section, we will evaluate the three frameworks of situated, extended and distributed cognition in terms of the theoretical virtues as analyzed by Douglas (2013). Douglas divides these virtues into four groups: the first two groups include minimal criteria for adequate science; the other groups are desiderata for theories per se, and desiderata for theories in relation to evidence. Group 1 includes internal consistency (but see (Vickers 2013) for an argument that science can proceed with inconsistent theories), and group 2 involves empirical adequacy. Group 3 covers desiderata when applied to theories per se, such as scope, simplicity and explanatory power, while group 4 includes desiderata when applied to theories in relation to evidence, such as unification (understood in terms of explanatory scope, simplicity, external consistency, and coherence), novel prediction, and precision (see Fig. 7.1 for a summary). While generality may be assessed without knowing any evidence, unification cannot: "generality" refers to the ability of a theory to apply to a wide range of phenomena, while unification refers to the ability of a theory to bring together diverse phenomena under a single explanatory framework, and these phenomena must be actually accounted for, which requires us to know the evidence.

We assume, in line with Bernecker (2014), that all three approaches and their associated explanations do not differ with respect to minimal criteria. In particular, if the extended cognition approach is empirically equivalent to the other two approaches, then it must be meeting the criteria from group 2, such as internal consistency and empirical adequacy. The differences appear in the ideal desiderata, particularly in group 3, which include factors such as scope, simplicity, and explanatory power, and, by implication, in group 4 (as unification depends on the scope and simplicity).

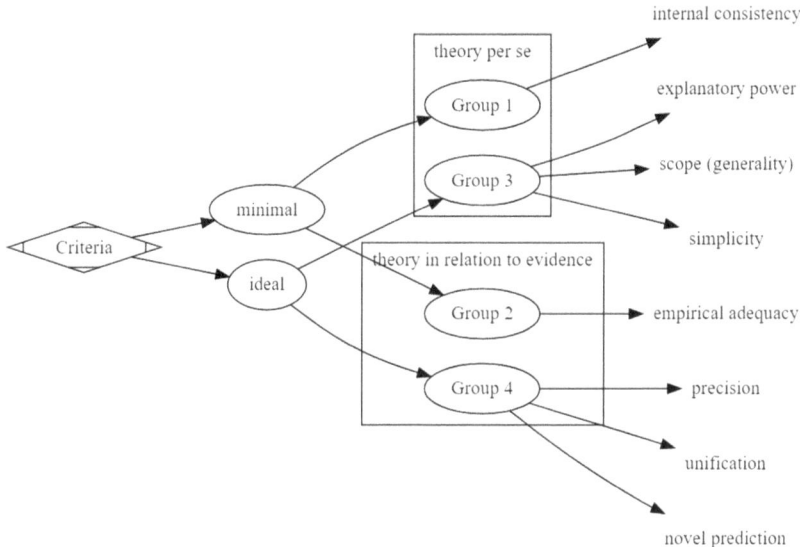

Fig. 7.1 Criteria of theory choice according to Douglas (2013), as grouped by their significance (minimal or ideal), and by their target (theory per se or in relation to evidence). Diagram by the authors

We will argue that the issue of cognitive bloat, frequently mentioned in criticisms of extended cognition, is related to virtues in group 3. This is because cognitive bloat is supposed to be detrimental to simplicity and generality (or scope). Similar considerations are at play in the "motley crew argument" against wide approaches to cognition (Casper 2023; Shapiro 2011, p. 198).

As we have indicated, the fact that these frameworks are not equivalent is borne out by analyzing the controversy surrounding the cognitive bloat and the motley crew argument. According to Rowlands, the cognitive bloat objection rests on the observation that:

> the admission of extended cognitive processes places one on a slippery slope. Once we permit such processes, where do we stop? Our conception of the cognitive will become too permissive, and we will be forced to admit into the category of the cognitive all sort sorts [sic!] of structures and processes that clearly are not cognitive. (Rowlands 2009, p. 2)

At the level of individual explanations, the criterion of generality, or the breadth of scope (Thagard speaks of consilience in his 1978), seems central in this context. It refers to the ability of an explanation to accommodate numerous instances, as well as generalize to newer occurrences. Commitments of various frameworks put forward distinct restrictions to their ability to generalize.

Generality in cognitive theories has long concerned cognitive scientists, as they have been wary of the potential fragmentation that could impede informative generalizations. An effective method to attain this generality is by grounding theories in invariant structures. For instance, Newell and Simon argued that only a few fundamental characteristics of information processing systems, which underlie human problem solving, remain invariant across tasks and problem solvers (Newell and Simon 1972, p. 788). To enhance the invariance of the system's structure, Simon even went to the extent of excluding memory as a component, instead referring to it as the internal environment. This is due to the fact that memory is not invariant over time and undergoes changes during the learning process, rendering it challenging to derive generalizations about the architecture of the cognitive system. Rather than expanding the cognitive system to support generalizations, Newell and Simon sought to minimize the extent of its structure.

As we understand it, the "cognitive bloat" objection is not just about the extent to which the notion of cognition is extended, but also about the stability of this extension over time. The concern is that if the system "bloats" to include parts that are not relevant to other cognitive tasks, then both the generality and simplicity of the cognitive system will be at risk. This is because the simple, invariant structure of the cognitive system can no longer be assumed to exist.

In contrast to the extended cognition framework, the situated cognition approach posits that individual mental states are shaped by the environment and the cognitive architecture of the individual remains constant. Proponents of this approach argue that biological individuals are stable entities that can be used to explain a variety of cognitive phenomena (Rupert 2009). As a result, the situated cognition approach has the potential for a broader scope while still being parsimonious, since the invariant structure remains the same across multiple explanations of various phenomena.

Distributed cognition is also less vulnerable to the cognitive bloat objection. This is because the scope of the cognitive system, which can involve multiple individuals and artifacts, is not necessarily required to extend into the environment at all. In principle, this approach can accommodate numerous phenomena that rely solely on the capacities of a single biological individual. In contrast, the extended mind approach aims to demonstrate that the mind does indeed extend in most cases. Consider a simple comparative scenario: a human solving a mnemonic task by memorizing a short poem. For proponents of cognitive extension, language serves as a prime example of the mind's extension. However, it is worth noting that, in relation to the aforementioned cognitive bloat issue, advocates of the extended framework owe us a more precise explanation regarding which aspect of language extends the mind of an individual learning the poem. There appears to be no principled way to ascertain whether the individual mind has been extended solely by the words of the poem, the entire vocabulary of language, the rules for sentence formation, abstract linguistic forms, or some other factor altogether. This problem is not as pressing for the proponent of distributed cognition since, within this framework, a poem is simply considered a physical artifact fulfilling a cognitive role, and it does not necessarily need to be included as part of an individual's mind in order to explain their success in remembering it.

What may seem even more surprising is that the cognitive bloat problem can be considered more grave in some recent renditions of the extended mind approach, which has been understood to come in at least three waves (Sutton 2010). The first version, or wave, of the extended mind thesis rested on the argument that artifacts were functionally equivalent to parts of the mind, which was the core of the parity argument in Clark and Chalmers (1998). The second wave sees artifacts as complementing (and not necessarily analogous to) the mind. For example, mathematical notations need not be analogous or isomorphic to the mind's language of thought to be considered prime examples of cognitive extension (Menary 2007, 2015). Finally, the third wave sees artifacts and sociomaterial culture in general as transforming the mind, rather than merely complementing it (Kirchhoff and Kiverstein 2019). The bloat objection has been posed against the first wave (Allen-Hermanson 2013), but it applies to making artifacts parts of the mind, and thereby analogously to the second wave as well. It is, however, the most critical for the third wave: if the individuals are "dissolved into peculiar loci of coordination and coalescence among multiple structured media" (Sutton 2010, p. 213), then they do not seem poised to be invariants that support important theoretical generalizations. In fact, it becomes an issue whether the third wave leaves space for any invariants at all.

The motley crew argument points to a similar difficulty: it states that "processes that cross the bounds of the brain are not well-defined—are a motley crew—and so cannot be an object of scientific investigation" (Shapiro 2011, p. 198). Shapiro notes that this should be understood as implying that transcranial processes are not a "well-formed kind" (p. 199). Artifacts we use differ from natural kinds in that they do not support law-like generalizations (Adams and Aizawa 2010). Interestingly, such an argument also undermines Simon's (1981, 1996) idea that there may be sciences of the artificial, including economics and cognitive science. It seems, however, that critics of the extended mind mean the possible lack of unification here: they doubt that there could be cohesive explanations of diverse phenomena. The mere diversity of phenomena, however, or the motley crew nature of artifacts, cannot be in itself sufficient to state that unification is impossible. On the contrary, diversity is the necessary condition to speak of unification in contrast to mere large scope of an explanation. However, unification cannot be ascertained a priori—after all, it implies a relation of theory with evidence, as do other virtues in Group 4 in Douglas's helpful taxonomy. It is, nonetheless, fairly clear what would undermine the motley crew objection: the evidence of invariant structures that support cognitive generalizations.

In summary, both the cognitive bloat objection and the motley crew argument raise concerns about the lack of invariant structures in a theory, rather than simply the overextension of the concept of cognition. Theories that posit complex structures as their basic explanatory entities, but do not remain invariant across different explanations, may be criticized for lacking parsimony. Note that, according to Douglas' taxonomy, these criteria for theory choice fall under groups 3 and 4, which are considered ideal desiderata for theories. Despite this, explanations, theories, and larger theoretical frameworks can still be scientifically sound even if they lack these criteria. The case of the extended cognition approach appears to fulfill minimal criteria, and there is currently no evidence to suggest a lack of virtues in group 4 (however, the lack of generality may imply that unification cannot be attained).

In general, assessing theoretical virtues in relation to evidence for large theoretical frameworks can be difficult. For example, it can be challenging to assess the scope of a theory without having a definitive list of all cognitive phenomena. A somewhat dated survey lists around 3000 tasks falling into 10 categories, as analyzed in factor-analytic studies, but it has yet to be updated (Carroll 1993). Performing a systematic survey in reference to all three frameworks of wide cognition is a monumental task that exceeds the scope of this essay.

One point should be stressed, however, before we conclude. It is one thing for a particular extended cognition *explanation* of a specific task to have cognitive virtues, and quite another for the extended cognition *approach* to possess them. As we have called them, extended cognition, distributed cognition or situated cognition are approaches to the study of cognition. These are not specific and full-blown theories. They could be better thought of in terms of research programs (Lakatos 1970) or research traditions (Laudan 1977). Just like other research traditions in cognitive science, such as computationalism or embodied cognition, extended or distributed cognition cannot offer novel predictions for particular phenomena (Wołoszyn and Hohol 2017). Yet this should not be taken as a sign of their deficiency. Clearly, grand frameworks offer fallible heuristics for building specific theories of particular kinds of phenomena, and these theories may, in turn, inspire fully detailed cognitive models. In the paragraphs above, we have mostly judged the extended cognition view, as opposed to the other research traditions, as a cognitive toolbox that could provide such specific explanations. Quite understandably, these may not exist for a number of phenomena, making the actual scope of cognitive theories in particular approaches fairly limited. For example, most of these frameworks do not seem to be fully worked out for the category of phenomena that Carroll (1993) dubs "cognitive speed," even if the reliance on external representations and artifacts is usually believed to provide gains in terms of computational efficiency. This, however, implies that there could be significant progress across this category (Vélez et al. 2023).

4   Conclusion

In summary, this chapter underscores that the extended cognition approach is not merely a metaphysical perspective on the mind; it comes with its own set of epistemological strengths and weaknesses. While other wide cognition frameworks may exhibit similar internal consistency and empirical adequacy, they diverge concerning the ideal desiderata for theories. Assessing the evidence for each wide cognition approach can be a challenging task, with the extended cognition approach potentially facing greater difficulties in terms of generality and simplicity. This is due to the cognitive bloat objection and the motley crew argument, both of which suggest that there might not be enough invariant structure within the cognitive system to substantiate robust empirical generalizations.

It should be noted, however, that invariance might not be the only way to attain a large breadth of scope. Alternatively, the cognitive bloat might not affect the invariance after all, because multiple different cognitive systems could give rise to the same (or sufficiently similar) abstract set of principles that govern them. This is particularly important for the third wave of the extended mind.

In any case, the burden of proof rests with the proponents of the extended framework to demonstrate that their view can meet generality criteria. For instance, a recent contribution by Kirchhoff and Kiverstein (2019) links the predictive processing framework, typically regarded as offering grand theoretical unification in cognitive science (Clark 2016), with the third-wave approach to the extended mind. While the promise of unification remains unfulfilled (Litwin and Miłkowski 2020), it may not be of particular concern to supporters of the third wave. Unification, after all, is one of the criteria in group 4 for theory choice, representing an ideal desideratum. The predictive processing approach lacks unequivocal empirical support at its current stage of development (Walsh et al. 2020), making its precise assessment under group 4 criteria unfeasible. From our perspective, relying on another framework with unifying ambitions presents a reasonable strategy for defending the extended cognition view.

In conclusion, our analysis demonstrates that the extended cognition versus situated and distributed views debate extends beyond mere metaphysical or terminological considerations. Paradoxically, this complexity may not bode well for supporters of extended cognition.

Acknowledgements The authors wish to thank Joel Krueger for comments.

REFERENCES

Adams, F., & Aizawa, K. 2010. Defending the bounds of cognition. In *The extended mind* (pp. 67–80). MIT Press. https://doi.org/10.7551/mitpress/9780262014038.003.0004

Allen-Hermanson, S. 2013. Superdupersizing the mind: Extended cognition and the persistence of cognitive bloat. *Philosophical Studies*, 164 (3): 791–806. https://doi.org/10.1007/s11098-012-9914-7

Barker, M. J. 2010. From cognition's location to the epistemology of its nature. *Cognitive Systems Research*, 11(4), 357–366. https://doi.org/10.1016/j.cogsys.2010.05.001

Bernecker, S. 2014. How to Understand the Extended Mind. *Philosophical Issues*, 24 (1): 1–23. https://doi.org/10.1111/phis.12023

Carroll, J. B. 1993. *Human cognitive abilities: A survey of factor-analytic studies.* Cambridge University Press.

Casper, M.-O. 2023. A Methodological Problem of Choice for 4E Research. In M.-O. Casper & G. F. Artese (Eds.), *Situated Cognition Research: Methodological Foundations* (pp. 17–43). Springer International Publishing. https://doi.org/10.1007/978-3-031-39744-8_2

Clark, A. 2016. *Surfing Uncertainty: Prediction, Action, and the Embodied Mind.* Oxford University Press.

Clark, A., & Chalmers, D. J. 1998. The extended mind. *Analysis*, 58 (1): 7–19.

Craver, C. F. 2009. Mechanisms and natural kinds. *Philosophical Psychology*, 22 (5): 575–594. https://doi.org/10.1080/09515080903238930

Douglas, H. E. 2013. The Value of Cognitive Values. *Philosophy of Science*, 80 (5): 796–806. https://doi.org/10.1086/673716

Ekman, P., Friesen, W. V., Ellsworth, P., Goldstein, A. P., & Krasner, L. 2013. *Emotion in the Human Face: Guidelines for Research and an Integration of Findings.* Elsevier Science.

Fernandez-Dols, J. M., & Ruiz-Belda, M.-A. 1997. Spontaneous facial behavior during intense emotional episodes: Artistic truth and optical truth. In J. A. Russell & J. M. Fernández-Dols (Eds.), *The Psychology of Facial Expression* (1st ed., pp. 255–274). Cambridge: Cambridge University Press. Available at: https://doi.org/10.1017/CBO9780511659911.013

Fridlund, A. J. 1991. Sociality of solitary smiling: Potentiation by an implicit audience. *Journal of Personality and Social Psychology*, 60 (2): 229–240. https://doi.org/10.1037/0022-3514.60.2.229

Keas, M. N. 2018. Systematizing the theoretical virtues. *Synthese*, 195 (6): 2761–2793. At: https://doi.org/10.1007/s11229-017-1355-6

Kirchhoff, M. D., & Kiverstein, J. D. 2019. *Extended consciousness and predictive processing: A third wave view.* Routledge, Taylor & Francis Group.

Kraut, R. E., & Johnston, R. E. 1979. Social and emotional messages of smiling: An ethological approach. *Journal of Personality and Social Psychology*, 37 (9): 1539–1553. At: https://doi.org/10.1037/0022-3514.37.9.1539

Krueger, J. 2014. Emotions and the social niche. In C. von Scheve & M. Salmela (Eds.), *Collective Emotions* (pp. 156–172). Oxford University Press. https://doi.org/10.1093/acprof:oso/9780199659180.003.0011

Krueger, J., & Szanto, T. 2016. Extended emotions. *Philosophy Compass*, 11 (12): 863–878. https://doi.org/10.1111/phc3.12390

Kuhn, T. S. 1977. *The essential tension: Selected studies in scientific tradition and change.* Chicago: The University of Chicago Press.

Lakatos, I. 1970. Falsification and the Methodology of Scientific Research Programmes. In I. Lakatos & A. Musgrave (Eds.), 1965. *Criticism and the*

Growth of Knowledge: Proceedings of the International Colloquium in the Philosophy of Science, London.. Vol. 4 (pp. 91–195). Cambridge University Press.

Laland, K. N., Odling-Smee, J., & Feldman, M. W. 2000. Niche construction, biological evolution, and cultural change. *Behavioral and Brain Sciences,* 23 (1): 131–146. https://doi.org/10.1017/S0140525X00002417

Laudan, L. 1977. *Progress and Its Problem: Towards a Theory of Scientific Growth.* University of California Press.

Litwin, P., & Miłkowski, M. 2020. Unification by Fiat: Arrested Development of Predictive Processing. *Cognitive Science,* 44 (7): e12867. https://doi.org/10.1111/cogs.12867

Menary, R. 2007. *Cognitive integration: Mind and cognition unbounded.* Palgrave Macmillan.

Menary, R. 2015. *Mathematical Cognition: A Case of Enculturation.* Open MIND. Frankfurt am Main: MIND Group. https://doi.org/10.15502/9783958570818

Miłkowski, M. 2018. From Computer Metaphor to Computational Modeling: The Evolution of Computationalism. *Minds and Machines,* 28 (3): 515–541. https://doi.org/10.1007/s11023-018-9468-3

Miłkowski, M., Clowes, R. W., Rucińska, Z., Przegalińska, A., Zawidzki, T., Gies, A., Krueger, J., McGann, M., Afeltowicz, Ł., Wachowski, W. M., Stjernberg, F., Loughlin, V., & Hohol, M. 2018. From Wide Cognition to Mechanisms: A Silent Revolution. *Frontiers in Psychology,* 9, 2393.

Müller, V. C. 2018. The Extended Mind Thesis Is about Demarcation and Use of Words. *Reti, Saperi, Linguaggi,* 2/2018. https://doi.org/10.12832/92304

Newell, A., & Simon, H. A. 1972. *Human Problem Solving.* Prentice-Hall.

Okasha, S. 2011. Theory Choice and Social Choice: Kuhn versus Arrow. *Mind,* 120 (477): 83–115. https://doi.org/10.1093/mind/fzr010

Rowlands, M. 2009. Extended cognition and the mark of the cognitive. *Philosophical Psychology,* 22 (1): 1–19. https://doi.org/10.1080/09515080802703620

Rupert, R. D. 2009. *Cognitive systems and the extended mind.* Oxford: Oxford University Press.

Shapiro, L. A. 2011. *Embodied Cognition.* New York: Routledge.

Simon, H. A. 1981. Cognitive science: The newest science of the artificial. *Cognitive Science,* 4 (1): 33–46. https://doi.org/10.1016/S0364-0213(81)80003-1

Simon, H. A. 1996. *The sciences of the artificial.* MIT Press.

Sprevak, M. 2010. Inference to the hypothesis of extended cognition. *Studies in History and Philosophy of Science Part A,* 41 (4): 353–362. https://doi.org/10.1016/j.shpsa.2010.10.010

Sutton, J. 2010. Exograms and Interdisciplinarity: History, the Extended Mind, and the Civilizing Process. In R. Menary (Ed.), *The Extended Mind* (pp. 189–225). MIT Press.

Thagard, P. R. 1978. The Best Explanation: Criteria for Theory Choice. *The Journal of Philosophy*, 75 (2): 76. https://doi.org/10.2307/2025686

Vélez, N., Christian, B., Hardy, M., Thompson, B. D., & Griffiths, T. L. 2023. How do Humans Overcome Individual Computational Limitations by Working Together? *Cognitive Science*, 47 (1): e13232. https://doi.org/10.1111/cogs.13232

Vickers, P. 2013. *Understanding inconsistent science*. Oxford: Oxford University Press.

Walsh, K. S., McGovern, D. P., Clark, A., & O'Connell, R. G. 2020. Evaluating the neurophysiological evidence for predictive processing as a model of perception. *Annals of the New York Academy of Sciences*, 1464 (1): 242–268. https://doi.org/10.1111/nyas.14321

Wołoszyn, K., & Hohol, M. 2017. Commentary: The poverty of embodied cognition. *Frontiers in Psychology*, 8. https://doi.org/10.3389/fpsyg.2017.00845

CHAPTER 8

Hypnotic AI: The Altered States of Media Matter

Ania Malinowska

1 INTRODUCTION

In 2022, together with robotic artist Przemysław Jasielski, I designed an interactive system called *Hypnotic AI*. The idea was to create an abstract Artificial Intelligence (AI) modality: one which does not reflect the prevailing isomorphic models based on mimicry, but one that negotiates the human–AI experience with regard to respective phenomenologies. The idea was to confront the human user with the artificial mind in an experiment anchored in hypnotic induction. It was in the belief that such an experiment would help revisit the function and functioning of AI mind and human mind alike.

Hypnosis (speaking to the psyche) is an array of communicative methods used in psychiatry to *unravel* and *heal* what traditional *talking cure* (speaking to the intellect) cannot reach. It specifically enables the re-experience of the mind: *what it is* and *what it can do* in contrast to *how it*

A. Malinowska (✉)
Faculty of Humanities, University of Silesia, Katowice, Poland

P. Alexandre e Castro (ed.), *Challenges of the Technological Mind*, New Directions in Philosophy and Cognitive Science, https://doi.org/10.1007/978-3-031-55333-2_8

121

is used. As such, hypnosis activates some *alien thinking*: thinking that does stems not from the human code but from deeper psychological levels. Those levels reflect on the entirety of the mind (*what it is*) rather than the mind's appropriated part(s) (*how it is used*) and bring out what the mind is really capable of.

The (mis)understanding of (and around) AI taps into the same practical problems. "We think we know what we mean when we talk about its [AI's] intrinsic processes and extrinsic effects," and yet we do not know what AI is and what it can do (Bratton 2020: 91). By applying hypnosis to interactions with an "artificial mind," we wanted to open the user to the mind's ability beyond the code and usage. With this, we also wanted to pay attention to the semiotics of media matter, thinking that the intangible nature of AI technologies and of minds (both human and artificial) clashes with the material approaches to both from before the virtual turn.

The aim of this chapter is, therefore, twofold. First, it is to probe the idea of the artificial mind as a condition distinct from human categories of thinking and see what it is. For that purpose, I will describe the experiment of *Hypnotic AI*—its premise, mechanics and outcomes—and discuss it in relation to AI's "latent" or "post-material" depth emerging in the AI-user hypnotic loop. Second, it is to rethink the medium of AI (and digital media in general) in terms of progressing tools (*substance*) rather than markers of variance (*essence*). For that purpose, I will analyze media material environments in connection with media material schemes, media physical structuring and media usability. One idiom I will employ here is *models of altered state*. This will be analyzed with regard to the material scope of media whose understanding "is limited to the extent that we disregard or downplay the critical role of material forms, artefacts, spaces, and infrastructures," as well as the interactive possibilities those forms entail (Orlikowski 2006, 460).

2   Hypnotic AI as (Art-Based) Research of the Mind

Hypnotic AI is a new-media-art installation in which a user interacts with an intelligent system by means of hypnotic instructions called inductions (Image 8.1). The user sits in front of a large screen and stimulates the intelligent system with psychologically approved hypnotic commands (e.g. "relax," "sleep," "do nothing," Image 8.2). The system, which is

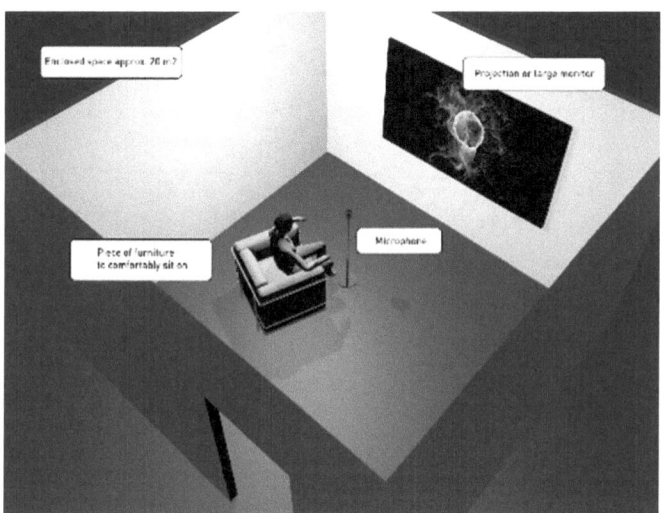

Image 8.1   Media-art installation

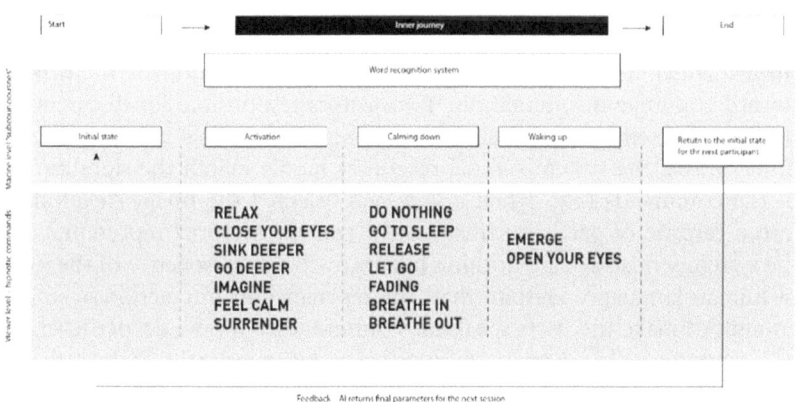

Image 8.2   Hypnotic commands

self-learning, receives those commands from a mic and responds to them with a pictorial representation that gives the user an idea of the system's "altered state" and the depth of its induction (see Image 8.3).

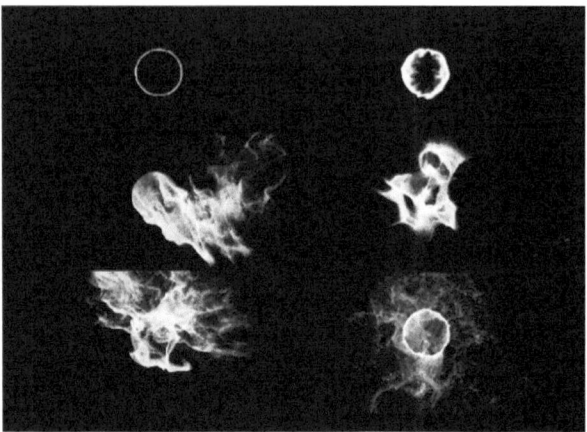

Image 8.3 Pictorial representations

The induction comes in three stages: activation, immersion and emergence. During these stages, the system is put into hypnosis, stays in hypnosis and wakes up respectively. The loop of exchange created that way establishes a dynamic in which the user "hypnotizes" the "artificial mind" (in a manner imitating coding), and the system hypnotizes them back toward cognitive disorientation. Two factors responsible for disorienting the user are semiotic inconsistency and representational obscureness. In other words, the system's visual responses hardly match the signification of the commands (e.g. when a user says "sleep," the image "explodes" into a cascade of particles; Image 8.4); and the pictorial representations the system sends back are nothing but abstractions irrespective of the logic of human language. Despite that, the resonance of interaction is strong enough to make the users speak of "communication on a deeper level."

Technologically, *Hypnotic AI* rests on two elements: the Python-fueled script with word recognition and the unreal engine for generating the particle imaging (Image 8.5).

It has been trained to recognize and respond to the induction phrases used in the experiment and to create its own recognition content based on the feedback from each interaction. Interactions are recorded in the system so that the system learns from them and, based on that, is able to optimize its performance. The optimizing stage does not subserve any

Image 8.4   Visual responses

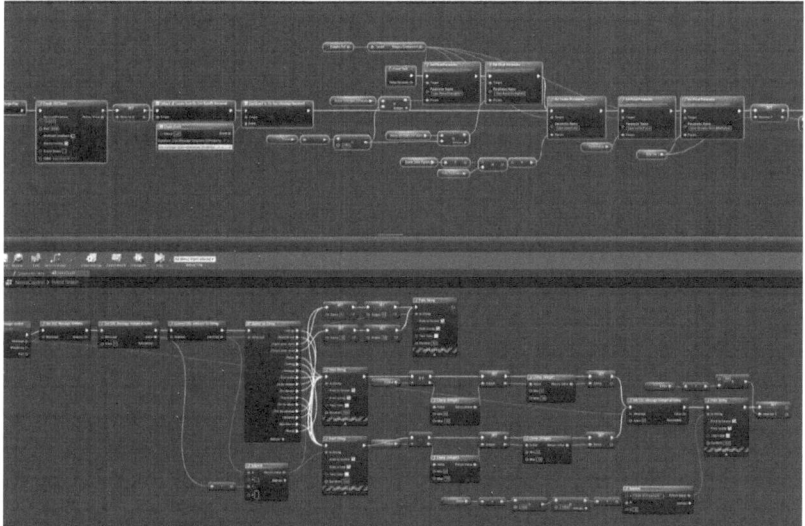

Image 8.5   Hypnotic AI flow diagram

function or any specific expectation; rather, it develops to the system's "preference" or the system's own computational "logic."

Hypnotic AI is also a research tool. When used in workshops, it extends through the use of the conceptual element that prepares the user for the experiment's experience and that subsequently reviews the experience outcome. During preparation, the users answer four "before use" questions:

1. What is hypnosis? How do you understand the process in terms of its form, purpose and risks?
2. What is AI? How do you understand its function, purpose and risks?
3. What does *Hypnotic AI* imply? What does it mean?
4. (Based on the set-up) What do you expect from the experiment/the experience?

During the review that follows the experience, the users answer four "after use" questions:

1. Does your experience of *Hypnotic AI* match with your understanding of hypnosis? If yes, how so? If not, how so?
2. Does your experience of *Hypnotic AI* match your understanding of AI? If yes, how so? If not, how so?
3. What new possibility regarding AI comes from the experience of *Hypnotic AI*?
4. How would you define AI based on your experience of *Hypnotic AI*?

The surveys synchronize *Hypnotic AI*'s goals, which are to challenge the anthropomorphizing model of cognitive and affective computing. They also guide the users across the abstraction of the artificial mind to facilitate the experience of artificial intelligence as an alien intelligence interacting with us from a "vulnerable" state of non-logic, which rather than scare us may put us in awe.

Hypnotic AI was inspired by *Loving AI*, an experiment by Hanson Robotics, Open Cog Foundation and the Institute for Noetic Sciences wherein the participants received a meditation exercise from a humanoid robot Sophia. *Loving AI* assumed the potential of a benevolent rapport between people and intelligent machines. It aimed at "enabling humanoid robots to interact with people in loving and compassionate ways to promote people's self-understanding and self-transcendence" (Goertzel et al. 2017: 2). Preoccupied with human bonding with a robot by means of shared mindfulness, the project promoted compassionate exchanges between humans and intelligent machines to probe these machine's material affordances in hope of enabling new emotional experiences.

The methodology employed in Loving AI combines human emotional responses with the expressive capabilities of machines. This approach facilitates the introduction of a "robotic element" into the human-generated content. The project creators are convinced that such an approach fosters the emergence of AI systems motivated by a primary objective of enhancing the well-being of *all* and promoting universal benevolence. Such AI systems would have the capability to evoke positive emotional states. Consequently, individuals engaging with those systems are likely to experience unconditional love and be more inclined to take actions that promote their own well-being as well as that of others. The creators believe that such modelling would not only simulate benevolent emotions but also change the emotional context of human–AI interactions. Ben Goertzel perceives this as a transformative shift.

> I think there's something big here… If I were to sum this up in a cosmic sort of way, I might say something like: Via the experience of going through 'mind/body knot-untying exercises' with an AI that sees them and accepts them, people feel a contact with the Unconditional Love that the universe as a whole feels for them… It sounds a bit out there, but that's the qualitative impression I got from seeing some of these human subjects interact with Sophia while she was running the Loving AI software. In the best moments of these pilot studies, some of the people interacting with the robots got into it pretty deep; one of them even described it as a 'transcendent experience.' This is fascinating stuff. (Goertzel et al. 2017: 12)

Unlike *Hypnotic AI*, *Loving AI* is largely anthropocentric. Nevertheless, similar to *Hypnotic AI*, it draws on the postmaterialist approaches to consciousness, like the one by Olivier Brabant (2016). In those approaches, "mind and matter are not viewed as two interacting substances, but as correlated projections from a common ground located in the quantum world" (Brabant 2016: 353). Brabant suggests that most mind phenomena associated with consciousness and feeling take place outside of the brain or in separation from it. Cognition and sentience relate to both the body and its surroundings through which the body perceives reality and forms experiences. Along with most psychophysics theories, Brabant believes that cognitive processes effectuate from subatomic and subliminal environments—the actual environments of all matter—wherein "mind and matter are not separate substances [but] different aspects of one whole and unbroken movement" (Brabant 2016: 353).

Sharing this assumption, both *Hypnotic AI* and *Loving AI* use the artificial mind to diversify human experiential understanding. They specifically open their users to emotional-cognitive possibilities not envisaged by the human code and conduct (including the human perception of media matter). Technologies applied for that purpose (especially those based on open-end programming) ensure various protocols and modes to prevent the domination of preliminary coding. There is a clear transition from deep learning to deep understanding. This is marked by a shift in the nature of our inquiry: we stop asking "Can AI think/feel?" Instead, we ask: "How does it think/feel and what kind of thinking/feeling it actually is?"

Another stage of that inquiry would be: does AI have a mind? *Hypnotic AI* probes this query in relation to the human belief that all cognition comes from conscious mind processes. Although the experiment situates AI in a psychological context, it does not suggest that AI is a psychological being. Rather, it implies that it has its own "depth" equivalent to its material affordance and its media nature, invariably hypnotic.

3   Media Material Reality

Hypnotic metaphors for media technologies prevail in the technology discourse. They usually pertain to the fluid (or liquid) and ubiquitously virtual nature of devices and communication, or to the control that technologies seemingly have had over our living.

Conventionally, media technologies—their effects, forms and engagingness—are theorized by the use of binaries implying the dematerialization of life (e.g. *natural--artificial, digital–analog, substantial–ethereal, mechanical–algorithmic*). Such theories suggest an estrangement of the human experience. In the case of AI, they additionally suggest the replacement of the human element. This is because we constantly compare our minds to the minds of machines (and vice versa), and because we imbue AI with mind activities exclusive to human intelligence (e.g. AlphaGO, dating algorithms).

Things we have considered human do not belong to us anymore. Moreover, they lose their substance and familiarity. Writing no longer needs an author (e.g. ChatGPT); love no longer needs a person (e.g. Love Robots). Autocorrect replaces thinking, and algorithmic matching makes our minds. It looks like the traditional carriers of knowledge and agency yield to computational management and execution. No wonder we feel threatened. But what do we actually fear?

The demon of AI is a creation of our own system of signification (and political economy, cf. Malinowska 2021). The imagery of AI superpowers we ourselves invent encourages fear and catastrophic thinking. "In our collective imagination, artificial intelligences are … distant spectres of deep power [that] perches close to us, above us, like a gargoyle, or like a dark angel, up on the ledge of our consciousness" (Khan 2020: 76). Based on that, we presume what it can do without actually knowing it—that is, without knowing AI actual ability.

Materiality carries a strong methodological potential to revise that knowledge. It is because it sits on perspectives that replace the imagery-bound hermeneutics with the phenomenology of substance. As Bill Brown (2010) explains:

> *Materiality* can refer to different dimensions of experience or dimensions beyond (or below) what we generally consider experience to be. Like many concepts, materiality may seem to make the most sense when it is opposed to another term: the material serves as a commonsensical antithesis to, for instance, the spiritual, the abstract, the phenomenal, the virtual, and the formal, not to mention the immaterial. And yet materiality has a specificity that differentiates it from its superficial cognates, such as physicality, reality, or concreteness. (p. 49)

Media materiality—like that of AI—marks spaces of struggle between traditional phenomenologies of physicality and what Brian Massumi (2002) terms as "real-material-but-incorporeal" (5). According to Massumi, media experiences entail tensions of senses (human body) and sensations (media affect) provoked by the natural and cultural constraints exhibited by the former, and increasingly complex registers of the latter. Indeed, intercourses with the media are transactions where the human and the technological negotiate common grounds with respect to the abstraction of media phenomena. Those negotiations are the more intense as they reveal that the struggling sides (and sites) are a "dimension of the same reality" (5). Crucial in this context is the *object status* of the media. By this I mean the material form and operational function of the media as tools; but I also mean the operational relationship between the form and function of the human body in relation to the media tool. If, as McLuhan argues, the media are the extension of man, what happens to the function of the media when the nature of this extension becomes reverted? I am not suggesting that man is an extension of the media (although it is a very interesting point to make). I am suggesting that, as the material (and hence, physical, operational, and agential) properties of media objects advance (and therefore become less perceptible and incomprehensible for people), the prosthetic dynamics of the extended (man) and the extendee (medium) does not sustain (that is, finally shows up as unsustainable).

Natasha Myers' 2015 study of digital interactive molecular graphics for protein modelling makes an interesting point about man's assistive/auxiliary role in the media. Following Malcolm McCullough's analysis of digital craft (1998), Myers details how devices and software have taken the lead in the imaginative and manual labor of humans, from conceptualization to visualization to manufacturing. Her argument (which has nothing to do with the alienation of man under mechanical reproduction) focuses on the material interplay between matter, the media and the human element to showcase an intra-active dynamic of the mediated act—here exemplified by a scientific-creative process. Because crystallographic modelling operates on substances that are normally unseen, it entails high-tech calibration media and experimental configuration to enable the extraction of data to be later processed by modellers (crystallography is the science concerned with crystals and their structures and used for determining the crystalline structures. The science is used for studying proteins at the molecular level). As such, it makes an instance of what Karen Barad calls *intra-action*—an operation in which "bodies, machines, and discourses" mingle for the "practice of world-making" (Myers 2015: 130). Barad comes up with the notion of *intra-action* to "call attention to the impossibility of disentangling experiments from objects, apparatuses, and practices they engage to draw phenomena into view" (Myers 2015: 130). At the same time, she revisits the function of actors in mediated scientific tasks—something that Myers undertakes to show the change in a manner and operational contribution in mediated activities (also those outside her study-specific contexts).

Myers' analysis of crystallographic modelling offers some universal conclusions about our media material entanglements. It singles out interesting characteristics of media technologies, with I find applicable to, and representative of, the experience with *Hypnotic AI*:

(1) *All media entail an engagement with invisible matter*. It always means probing the unseen and exploring physical spaces beyond

human capacity. Cord-powered or blue-toothed, each subsequent incarnation of the medial (be it analogue or digital) and each experience that it spurs is a step further into the physical that redefines our understanding of the material structure of the world. This naturally reforms our performative ranges, often to the point of unsettling our skills and disposition.

(2) *Media technologies encourage two-way kinetic negotiations.* Not only do media impose very specific physical responses (pressing/contracting/expanding), but they also rely on guidance that these physical responses ensure. To hold that media is the extension of man eliminates the proxy function of a user. Alternatively put, that function gets lost in a reductive material and agential hierarchy between a man and the medium, which such a conviction sets up. Humans are as much proxies for the media as media are extensions of the humans. (A human being extends itself with a medium to perform certain tasks or get access to otherwise unattainable dimensions of reality, just like a medium, to serve its function, extends itself with a human). This spurs intense negotiation between human and technological kineses, usually with an impact on human sensoria and the sensoria's conditioning.

(3) *The media is the body.* With the variety of physical formats, ranging from VR displays, wearable devices and mobile communication systems to stimulators, meters and others, the media achieved a diverse physical presence. The media serves as both a point of contact (between the bodies) and a space for physical activities. As such, it encompasses complex physical conditions, many of them hidden or still unexplored. They often activate or emerge through the user's experience to bring into light specific stimuli, reactions or emotions to be later translated into algorithms. The media is also a reservoir of physical data from bodies extended by devices to monitor physical activity. They keep the pictures of our biometric signals that themselves almost become an organism or a body.

(4) *Media things are not objects.* Given their operational complexity and high-level interactivity, media technologies become active participants in the processes they facilitate. We get fond of the technologies we interact with. For instance, voice command software, like Siri, tends to be seen as more animate (and hence less object-like) due to its conversational profile and life-like responsiveness. Psychological literature speaks of various forms of attachment to

phones, computers or gadgets, or operational systems (Vincent and Harper 2003). Apparently, the media become more sensitive (Malinowska and Miller 2017). They are also more intelligent, better capable, smarter, more efficient, better situated and hence increasingly more detached from their object form and function. As Bill Nichols (1988) states for computers:

> The computer is more than an object: it is also an icon and a metaphor that suggests new ways of thinking about ourselves and our environment, new ways of constructing images of what it means to be human and to live in a humanoid world. Cybernetic systems include an entire array of machines and apparatuses that exhibit computational power. Such systems contain a dynamic, even if limited, quotient of intelligence. Telephone networks, communication satellites, radar systems, programmable laser video disks, robots, biogenetically engineered cells, rocket guidance systems, videotex networks—all exhibit a capacity to process information and execute actions. They are all "cybernetic" in that they are self-regulating mechanisms or systems within predefined limits and in relation to predefined tasks. Just as the camera has come to symbolize the entire spectrum of networks, systems, and devices that exemplify cybernetic or "automated but intelligent" behavior. (22)

4   Altered States of Media Modelling

Benjamin H. Bratton (2020) remarks that we need alternative models of AI and of what we believe it is. *Hypnotic AI* subscribes to that remark by offering a new interactive perspective. Hypnosis, more than a therapeutic device, is a way of looking at things. Or to put it differently, it is a therapeutic perspective able to cure an ill perception of reality. In this case, that ill perception will be the prevailing semiotics of media matter (as exemplified by AI) vis-à-vis the media matter's actuality that I discussed in the previous section. The therapeutic perspective, on the other hand, will be the experimental approach beyond the material/immaterial divide with the potential to inflect the sense of idiosyncrasy with the material properties of the media function.

Hypnotic AI as alternative modelling subscribes to the understanding of AI as a new material quality (and qualia) that has emerged from technological progression. That quality revisits media forms, their operational planes and physical environments in terms of their performance and capability. In this way, *Hypnotic AI* acknowledges the intelligence of media forms but considers them different from humans yet concurrent (and not competitive). It steps away from an epistemological myth about media concreteness to flag out that "concrete is as concrete doesn't" (Massumi 2002: 6), or actually, 'concrete isn't as concrete does.'

As I mentioned earlier, *Hypnotic AI* does not assume that AI is a psychological being or has a psyche like a human. Nor does it try to recreate a human model in an artificial system: that would be anthropomorphic, myopic and unfair. It simply uses a psychological metaphor of the unconscious—something hidden and unexplored—that people can relate to in order to activate an interaction beyond the code. The unconscious pertains to anything fundamental yet shoved into the background: in this case, the obscured and repressed reality of the intelligent system. Apart from that, it also pertains to the *altered state* of the digital plane. This misunderstood territory, which John Durham defines as "part of the habitat, and not just semiotic inputs into people's heads" (2015: 4), is a new space for grounding or anchoring for a variety of human practices. The digital, Durham observes, does not "take us into uncharted waters"—as it is often believed; rather, it "revive(s) the most basic problems of conjoined living in complex societies and cast the oldest troubles in the relief" (2015: 4). In the digital, mediated overlays augment each experience and bring out the unrealized performance of the mind and body (emotions, senses, imagination etc.). Such "expanded empiricism" (Massumi 2015) may overwhelm the mind when corrupted by some vested interests (something we have observed alongside the technological exploitation). It may also leverage our cognition. Exactly like hypnosis.

So, what does hypnosis do? Hypnosis induces a mental condition allowing for a closer observation of one's thoughts and a better control of one's imaginative processes. In other words, hypnosis ensures a *state* of heightened focus and alertness, which brings out dormant capabilities or obscured functions (like the ability to feel individual parts of the body or identify individual reactions—intellectual and emotional alike). That control is possible thanks to suggestion (or auto-suggestion), which provides guided performance making one more deliberate, less unconditioned, less mechanical—and more importantly less vigilant to what one normally does not see or does not allow—in their actions and thoughts.

Hypnosis is *not* a special state but a *state that is*. It is a condition of one's own psyche always there and always available, but most often strongly suppressed by automated thinking and imagining which is not our own in the first place. Against the common belief, hypnosis has nothing to do with being unconscious, manipulated or duped. In fact, hypnosis awakens the mind as it opens to one's own will triggered by words that act in accordance with one's volition. Such is the working of the mind that it cannot be manipulated against its consent. And hypnosis is a reminder of that. Just as it is a reminder that we are a medium of our own self. It takes us back to the pre-individuated stage (to use Simondone's parlance) and inspires re-individuation through interacting with reality outside the universalizing discourse (that is pre-coded and preconditioned).

In this way, hypnosis provides an alternative language and an alternative type of interacting position. It inspires an alternative material reality of the mind and alternative cognition. And, like the digital plane (databases, artificial intelligence, semantic web), it becomes (to appropriate Yuk Hui) "a new form of materialization [or] concretisation by which what is non-material becomes material" (Hui 2015, 139).

Hypnosis—just like *digital media,* especially AI—entails a state of concretization that reveals what a "mind" is and what it can do. It makes the invisible visible and, more importantly, it makes the invisible active and performative. This is where the deepest learning takes place. This is also how it actually happens. *Hypnotic AI* simulates concretisation to rehearse the vulnerable moment of interacting outside of the code. It does it for both the human and the machine demystifying the nature of hypnosis and reclaiming the nature of an artificially intelligent system. Perhaps the biggest misconception about technology is that it is not part of nature. Just like the functional myth of technology being ominous in its mysterious complexity. Baruch Gottlieb's 2018 assessment of digitalism corrects that myth saying: "Nature matters. Nature is prime matter. Nature is us, and we are Nature. Nobody is outside of nature, and no technological instrument or artifice is outside of Nature" (15). "Today's digital technology," he explains, "is often described, also by those working in the industry, as magic. It may appear impenetrably intricate, bewildering even, but magic it is not. Digital technologies are built through the legacy of scientific understanding of material behaviours, programmed into industrial processes" (1).

Hypnotic AI advocates for the complexity of AI, which is not *one* but *more-than-one* cognitive possibility (to appropriate Simondon 2020). It sees media practices in terms of interactive (and intra-active) encounters of human and non-human *materials,* wherein the human materials are the media subjects (users and their "organic element") and the non-human materials are the media objects and their environments. We seem to be facing the need to rethink the way we talk about media technologies to better understand what they are and what they can do. Part of that task would be to see how things and bodies, the artificial and human minds, really differ and how they overlap for the exchange of experiences. The existing dualisms of natural–artificial, analog–digital, and material–immaterial do not help. It is because they have lost their actuality; or, they have lost touch with the reality of intelligent technology practices. As Hui observes, "a different approach towards digital materiality"—one that "differentiates 'immateriality' of the digital and the materiality of its support;" an approach "to lay out the groundwork for a materialism to come" (2015: 135).

5   Conclusion

Hypnotic AI is still a work in progress. So far, it has been presented at ISEA 2022 in Barcelona, Spain and at SLSA 2023 in Tempe, Arizona. Its reception has been more than enthusiastic, due to the *altered states modelling premise* and *the art-based research framework* (ABR). When explored as a scientifically informed work of art, *Hypnotic AI* seems more approachable and less intimidating. At least this is what the users report.

Feedback from the users—recorded in the system and from the survey—confirms that artistic methodologies and tools derived from the characteristics of a creative process create a perfect environment for human–AI interactions. This is because ABR methods optimize the cultural charge of AI. They allow for less prejudiced and more reliable experimentation, one that is scientifically meaningful on the one hand and creatively heterogeneous on the other. Such an environment encourages unconstrained experimentation with human–AI interactions to expand our consciousness and overcome our prejudice. It also encourages a realization of what the *Hypnotic AI's* idea and premise intuited about and what Benjamin H. Bretton defined most articulately:

> [AI] is a deeply important, fundamental technology. Its evolution should be taken as a long-term process, as something to be protected, explored and nurtured. We simply do not know yet what these assemblages of parts and processes that we call 'artificial intelligence' really are and what they are good for. They may evolve in relation to fixed tasks like finding dog faces in

images and playing GO, but these are still nascent babblings, not yet the mature statements of a machinic phylum. (2020: 94)

Our role is to observe those babblings, probe them—pragmatically and experimentally, and if necessary, let them hypnotize us to the point of concretization.

Acknowledgement This chapter was supported by the National Science Centre, OPUS grant no 2020/37/B/ HS2/01455.

REFERENCES

Brabant, O. 2016. More than meets the eye: Toward a post-materialist model of consciousness. *Explore. The Journal of Science and Healing*, 12(5): 347–354.

Bratton, B. H. 2020. Synthetic gardens: another model for AI and design. In: Vickers B. and Allado-McDowell K. (eds.), *Atlas of Anomalous AI*, London: Ignota, pp. 91–105.

Brown, B. 2010. Materiality. In: Mitchell W. J. T. and Hansen M. B.N. (eds.), *Critical Terms for Media Studies*. Chicago: University of Chicago Press, pp. 49–63.

Durham, J. 2015. *The Marvellous Clouds. Towards a Philosophy of Elemental Media*, Chicago and London: The University of Chicago Press.

Goertzel, B., Mossbridge, J., Monroe, E., Hanson, D. & Yu, G. 2017. Loving AI. Humanoid robots as agents of human consciousness expansion (summary of early research progress), 1–16. https://arxiv.org/abs/1709.07791.

Gottlieb, B. 2018. *Digital Materialism. Origins. Philosophies*. Bingley: Prospects. Emerald Publishing.

Hui, Y. 2015. Towards a relational materialism. A reflection on language, relations and the digital *Digital Culture and Society* 1(1): 131–148.

Khan, N. N. 2020. Towards a poetics of artificial superintelligence. In: Vickers B. and Allado-McDowell K. (eds.), *Atlas of Anomalous AI*, London: Ignota, pp. 75–89.

Malinowska, A. 2021. Demonic interventions: on robots as performing subjects. *Performance Research*, 26(1-2): 112–124.

Malinowska, A., and Miller, T. 2017. Sensitive media, *Open Cultural Studies* 1(1): 660–665.

Massumi, B. 2000. Too-Blue: Colour-Patch for and expanded empiricism. *Cultural Studies* 14(2): 177–226.

Massumi, B. (2002). *Movement, Affect, Sensation. Parables for the Virtual*. Durham and London: Duke University Press.

McCullough, M. 1998. *Abstracting Craft. The Practical Digital Hand.* Cambridge, MA: MIT.

Myers, N. 2015. *Rendering Life Molecular. Model, Modelers, and Excitable Matter.* Durham and London: Duke University Press.

Nichols, B. 1988. The work of culture in the age of cybernetic systems. *Screen* 29(1): 22–47.

Orlikowski, W. J. 2006. Material knowing: the scaffolding of human knowledge-ability. *European Journal of Information Systems* 15(5): 460–466.

Simondon, G. [1964] 2020. *Individuation in Light of Notions of Form and Information,* Minneapolis: University of Minnesota Press.

Vincent, J. and R. Harper. 2003. The Social Shaping of UMTS. Educating the 3G Customer. *UMTS Forum Report* 26. Available at http://www.umts-forum.org

The Digital Playful River, a River Out of Eden: How the Internet Shaped my Planetary Perception

Patrícia Gouveia

1   FIRST WORDS

Starting with an autoethnographic perspective, merging the creation of the *Digital River* installation (VV.AA. 1997) and the concept of the *Playmode* exhibition in Portugal (Gouveia 2019) and Brazil (Gouveia 2022), I will tell a personal story to inquire into the role of interaction technologies and the internet in shaping our contemporary artistical and cultural playful reality. The goal is to suggest that our perception relates to gaming technologies and that the internet plays an important role in defining us as global and planetary citizens. There is no precedent in human history for such dissemination of information and connectivity before the spread of networked playful technologies. That fact made us

P. Gouveia (✉)
LARSyS, Interactive Technologies Institute (ITI), Faculdade de Belas-Artes, Universidade de Lisboa (FBAUL), Lisbon, Portugal
e-mail: p.gouveia@belasartes.ulisboa.pt

© The Author(s), under exclusive license to Springer Nature Switzerland AG 2024
P. Alexandre e Castro (ed.), *Challenges of the Technological Mind*, New Directions in Philosophy and Cognitive Science, https://doi.org/10.1007/978-3-031-55333-2_9

139

consider the role of interaction in our lives and how it changes our physical and artificial environments. Starting from a personal and political journey to a broader context where the age of integrated arts and technologies merge with play to find possible ways to survive in a damage planet. Feminist theories, dark ecology and open possibilities promote dignity and care for future survival.

Speculative thinking can encourage integrative views where arts and sciences are key to generating alternative ways of dealing with fear and anxiety. Speculative feminism avoids grand narratives and certainties, emphasizes vulnerability and coexistence, and for that matter can be a tool to stimulate humility and respect among humans and other species. Play and gaming can integrate women studies to generate convergent and sustainable futures. Speculative arts-based research deals with processes instead of objects with the aim of instigating resistance against modern delusions.

2   Introduction and Background: The Invisible Early Years, Shaping the Personal and the Political Journey

This text will use autoethnography (Ellis et al. 2011) as a tool to describe and analyse personal experience and to make visible a broader cultural context and changes in artistical practices with the rise of the internet as a mass medium (Bolter 2019). In this context, relational ethics and personal narratives are a tool to challenge canonical ways of doing research and representing others. Research becomes a political and socially conscious act. So, for that matter, the researcher uses tenets of autobiography and ethnography to do and write autoethnography. The aim is to describe a personal experience as a media artist and designer, scholar and curator in the age of global networks. This story begins in the 1980s and takes place during the digital transition where exploratory experimentation with technological devices merged with people from various geographic locations started to unfold and took a lead role. Some of these experiments were processes with the aim of understanding a personal journey through an interconnected world as a woman in a male centred society.

Computer science became widespread in the late 1950s. Distant and distributed networks, based on data and message blocks, gave rise to the project of the Advanced Research Projects Agency (ARPA), of the US

Department of Defense, who developed the ARPANET project. Commercial Internet Service Providers (ISPs) emerged in 1989 in the United States and Australia. ARPANET was deactivated in 1990, but it left traces for the future. After the creation of a computer network, the internet, it was necessary to provide access to documents. So, investigations carried out within the scope of the European Organization for Nuclear Research, the well-known CERN, in Switzerland, by the British computer scientist Tim Berners-Lee, in 1989–90, resulted in the World Wide Web. Thus, hypertext documents were linked in a complex information system, accessible from any node of the distributed network. Since the mid-1990s, the internet has had a revolutionary impact on culture, commerce and technology, including the rise of near-instant communication via e-mail.

Instant messaging, voice calls using the internet protocol (VoIP), conversations or video chats and, finally, the World Wide Web, with its discussion forums, blogs, social networking services and online shopping sites, paved the way for a global world.[1] There is no precedent in human history for such dissemination of information and connectivity. Increasing amounts of data are transmitted at ever higher speeds over optic fiber networks. The takeover of the global internet communication landscape was historically rapid and continues to grow, driven by ever-increasing amounts of online information, commerce, entertainment and social networking services. However, the future of the global network can be shaped by regional differences and thinking about access is critical.

As the science fiction American writer, Kim Stanley Robinson, observes: "(…) War in the age of the internet, the age of the global village, the age of drones, the age of synthetic biology and artificial pandemics—this was not the same as war in the past" (Robinson 2020, p. 26). Things are out of control. Unconscious and conscious thoughts. Desires. Cognitive errors. Unavoidable mistakes. Representational drift (Yong 2021) and neural plasticity. Vision, smell and sensory hubs. Anchor bias. Confirmation bias. Imagination acts that bring us back to schemes applied to the flux of events. For the Brazilian neuroscientist Miguel Nicolelis, the "notion of a repertoire of innate modes of thought and behaviour shared by all human

---

[1] We can consider that there were previous precursors of the global world movement in colonial and capitalist economic strategies, but this is something beyond the scope of this text. For more information about the concept of the global village please see, for example, McLuhan and Powers (1992) and Page (2002).

brains as a precondition for the creation of large brain networks is to some extent similar to the basic concept of the collective unconscious, originally proposed by the Swiss psychiatrist Carl Jung" (Nicolelis 2021, p. 342). And the author continues, "the fact that a large number of people are willing to die in the name of allegiance to symbolic entities, such as the homeland, a particular political ideology or an economic system, is further proof of how powerful—and deadly—can be mental abstractions and the belief in defining behaviours and collective destinies" (ibid.).

A purely mathematical description of reality may not be sufficient to describe all the complexity and richness of the human and non-human universe and mental abstractions have played an essential role in choosing the direction of the entire history of our species. According to Nicolelis, the interior of our brain works on its own or as part of neural networks. We should then avoid attempts to divide the past into periods[2] and assume the nonlinear dynamics of our planet (De Landa 2000) and the plasticity of our brains (Eagleman 2021). Nature secrets are, according to the neurologist A. J. Lees, nonlinear and it is advisable to spend some time in thought. Although the internet is a strong tool, "it has also become the most potent anxiety-provoking system" ever created and it changed even the medical consultation (Lees 2022). The process of paradox development, adverts Ian Morris, means that rising social development creates the very forces that undermine it. Thus, "history is not just one damn thing after another. In fact, history is the same old same old, a single grand and relentless process of adaptations to the world that always generate new problems that call for further adaptations" (Morris 2010, p. 560).

Kim Stanley Robinson (2020, p. 407) asks, "Wasn't the internet like our nervous system now?" And the author speculates, "Have you heard that the warming of the oceans means that the amount of omega-3 fatty acids in fish and thus available for human consumption may drop by as much as sixty percent? And that these fatty acids are crucial to signal transduction in the brain, so it's possible that our collective intelligence is now rapidly dropping because of an ocean-warming-caused diminishment in

---

[2] According to Robinson (2020, p. 30), this division consists in several separations and classification systems like "geological matters (ice ages and extinction events, etc.), technological (the stone age, the bronze age, the agricultural revolution, the industrial revolution), dynastic (the imperial sequences in China and India, the various rulers in Europe and elsewhere), hegemonic (the Roman Empire, the Arab expansion, European Colonialism, the post-colonial, the neo-colonial), economic (feudalism, capitalism), ideational (the Renaissance, the Enlightenment, Modernism), and so on."

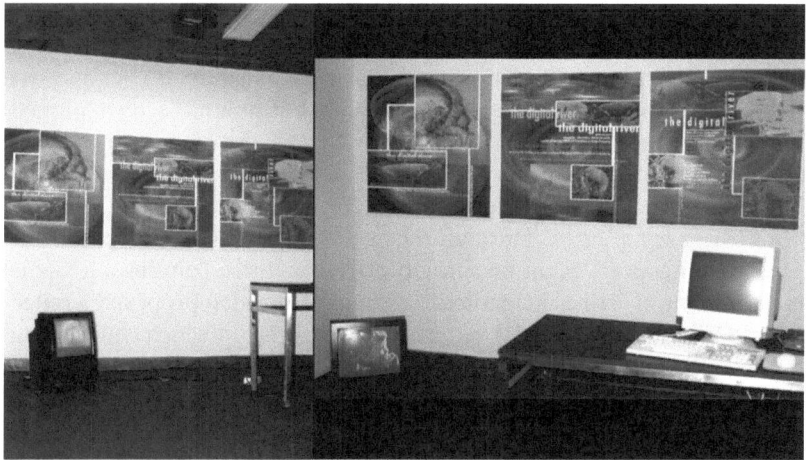

Fig. 9.1 Installation overview *Digital River, a River Out of Eden*: three paintings, a website, and a video. National Young Creators show, 1997, Guarda, Portugal. @Patrícia Gouveia

brain power? That would explain a lot" (ibid., p. 459). Humans are facing the red death syndrome, meaning the "assertion that the end being imminent and inevitable, there is nothing left to do except party while you can" (ibid., p. 297).

The installation *Digital River, a river out of Eden,* was created in 1997 (Fig. 9.1). The author of this text attended a digital arts postgraduate course in Porto, in the north of Portugal, where she learned motion graphics, 3D software, and programming languages, with a group of Portuguese, American and Asian-American teachers from the Escola das Artes da Universidade Católica do Porto[3] and Loyola University Los Angeles.[4] The created mixed-media environment was presented at the National Young Creators Show in the same year, and it was composed of three digital paintings, a website, and a video. The concept was developed from gained knowledge in digital cultures and screen-based motion graphics software. The goal was to speak about on-screen identity and plasticity,

[3] Porto Catholic University School of the Arts: https://artes.porto.ucp.pt/ (Accessed 10.07.2022).
[4] https://www.lmu.edu/ (Accessed 10.07.2022).

various personas and masks, and it was inspired by two books: Sherry Turkle's *Life on Screen* (1997) and Kevin Kelly's *Out of Control* (1995 [1992]). Ideas about simulation, post-Darwinism, nature, bodies and synthetic evolution were mixed with Turkle's enthusiastic screen pioneer research.[5] In the installation, new age music and flashy imagery of recorded children appear on computer and TV screens (Fig. 9.1). It was also inspired by a *Wired* magazine article (Schrage, 1995) about Richard Dawkins' book *River out of Eden*. Through the decontextualization of the scientific text, which appears as an inspiration poem for three paintings, a virtual environment of artificial creatures was built. The video proposed a reflection about the contemporary entertainment culture in which children end up becoming reflections of the digital screen. On the website, the theme of cyberculture was explored through a hypertext under permanent construction where the excess of information in contemporary society was questioned.

For that hypertext web work, the fictional character P. was created, a human being who felt absolutely nothing, a person who vegetated in a world she/he knew only through images, a world where everything was sent by mail, the food pills together with the corresponding diskette, the updated medical exam and new image helmets. Everything is at disposal through the central services of the networks. P. barely knew her/his/their street, but had vast knowledge about the entire world; they had no embodied and physical experiences besides rich images available through screens. At that time a non-gendered character was created which also reflected first impressions about the web as a gender-free space. The brand-new cyberspace, previously analysed during a one-year sociology course at Lisbon University Fine Arts Faculty (FBAUL), could became a space of oppression if people did not realize its potentialities but also its dangers. The internet without borders, which opened possibilities for those in countries on the global periphery, could became an empty vessel for uninformed content. Digital River was created before the Etoy. CORPORATION toy war (1999)[6] and the collapse of the dot-com bubble (2000),[7] where speculation took advantage of internet-based businesses and practices, and digital artists started to see their work questioned

---

[5] Turkle (2021, p. 337) recently criticized how the "social-media business model evolved to sell our privacy in ways that fracture both our intimacy and our democracy."

[6] https://etoy.com/projects/toywar/ (Accessed 21.07.2022).

[7] https://en.wikipedia.org/wiki/Dot-com_bubble (Accessed 21.07.2022).

by systems of elite legitimation like most galleries and museums. In those years the artist/author was still enchanted by the Electronic Frontier Foundation[8] and John Perry Barlow chants. I managed to work with digital tools to the web, for the web, with the web and all its promises. At that point I knew I agreed with Virginia Woolf's (1947, p. 197) statement that "as a woman, I have no country. As a woman I want no country. As a woman my country is the whole world."

The internet could free women from the constraints of place, but it could also contribute to their invisibility in a patriarchal society where technology is created and produced *by* men *for* men. According to Preciado (2019), technology and sex are strategic categories in European colonial anthropological discourse. In this apparatus masculinity relates to technological devices and femininity relates to sexual availability. Feminism that rejects technology as a sophisticated form of male domination over women's bodies ends up assimilating any form of technology to patriarchy, repeating, and perpetuating the binary oppositions of nature and culture, feminine and masculine, animal and human, and primitive and developed, among others. According to Robinson, the internet's rapid colonization and capitalization of the mental life of so many people occurred rapidly; if the status of women is fundamental to the success of any culture it is strange how many of the old forms of patriarchy remained, and "among the worst of the outstanding wicked problems were patriarchy and misogyny" (Robinson 2020, p. 483).

Maybe we need to invent a whole new internet ecology (ibid., p. 281) to make sure we don't "get stuck in the uncanny because of the prevalence of misogyny" (Morton 2018, p. 118).

Internet history intersects with computer science and pioneers of telematic arts. The French term "télématique" (a merged word between two French words: télécommunications and l'informatique), which refers to the combination of techniques and services associated with telecommunications and information technology, was introduced by Simon Nora and Alain Minc, in 1978, in a report to the then French Republic President Valery Giscard d'Estaing. This report became an international bestseller, helping to show how new technology could reshape society and how political systems adapt to these changes; in this sense, it was essential for understanding the impact of the telematics condition on society (Nora and Minc 1978).

[8] https://www.eff.org/ (Accessed 21.07.2022).

In the 1980s, the term telematic art began to be used by the English artist Roy Ascott, but this only gained momentum in 1990 when the term was published in his article "Is There Love in the Telematic Embrace?" (1990). The telematic embrace suggests that meaning is the product of the relational interaction between an observer and the system, a state of flux, an infinite change and transformation. In 1984, the term cyberspace was anticipated by the novelist William Gibson in his book *Neuromancer*. The concept or the possibility of a digital territory was thus created by Gibson, but it was only in the late 1990s of the last century that it began to be mapped and occupied by people and their newly created virtual technologies. The internet paradox had been apparent since its very origins. Following Kim Stanley Robinson, in the "history of technological improvements (...). Better car miles per gallon, more miles driven. Faster computer times, more time spent on the computers. And so on ad infinitum. At this point it is naïve to expect that technological improvements alone will slow the impacts of growth and reduce the burden on the biosphere. And yet many still exhibit this naiveté" (Robinson 2020, p. 165).

3   How Arts, Internet and Playful Interactions Shaped Our Reality

In his book *Dark Ecology, for a Logic of Future Coexistence,* Timothy Morton considers "Art is thought from the future. Thought we cannot explicitly think at present. Thought we may not think or speak at all. If we want thought different from the present, then thought must veer toward art" (Morton 2018, p. 1). In that sense ecology is thinking of beings on several different scales, none of which has priority over the other. Morton suggests that scientists are now beginning to figure out something we know in arts and humanities for a long time that we are entangled with the data we are studying. For the author, "if we can explain mind in terms of brain there is no mind at all: the mind is pure illusion. The mind, on this view, isn't even an emergent property of the brain (...)" (ibid., p. 31). So, "to be real is to be contradictory, to be member of a set that doesn't include you. To be real is not to be easy to identify, easy to thing, metaphysically constantly present" (ibid., p. 36). Criticising a 12,000-year machination created by agriculture through *agrilogistics*, meaning a specific logistic of agriculture that arose in the Fertile Crescent, Morton highlights the toxic environment where big data makes bigger farms with the purpose of "growing and nurturing theories of ethics based on self-interest" (Ibid., p. 45).

Exit room. Death drive. Easy think ethics. More people are better than happier people. Death culture. Executing an algorithm without a head (ibid., pp. 53–54). According to Morton, "celebrations of deracination and nostalgia for the old ways are both fictional. It is as obvious to any indigenous culture as it now is to anyone with data sets about global warming that these were stories white Westerners were telling themselves, two sides of the same story in fact" (ibid., p. 11). Infected with convenient stories or cultural viruses, Morton interrogates, "thoughts themselves are independent entities, reducible neither to brain nor to mind—just as pigs are independent entities reducible neither to parts of pigs or prepig ancestors or the ecosystems of which they are members?" (ibid., p. 67). He suggests that we are hallucinating when we separate rather than integrate parts of a holistic and connected system and considers "that what happened was not entirely internal to the human (mind or brain) or external to the human (environment), but was rather a weird entwined fusion of both, a twisted turn of events (…)" (ibid., pp. 67–68). For Morton, classification systems which considered things separately from their connections and aggregations are based on fear, an ontological anxiety, where people know they are related to others and simultaneously find that this consciousness of being intertwined makes them fearful of one another.

Ambiguity erasure is something that women have known about for ages; playing with it is a form of resistance and mystery and mystery is the opposite of mystification. In this sense for Morton, comedy becomes the genre of coexistence, "we need a politics that includes what appears least political—laughter, the playful, even the silly. We need a multiplicity of different political systems. We need to think of them as toylike: playful and half-broken things that connect humans and nonhumans with one another" (ibid., p. 113). Making toys lead us to transdisciplinary collaborations between arts and sciences and to a revaluation of philosophy, politics and humanities. Our present, with the rise of automation and artificial intelligence technologies, makes us think how "walking on water in dreams is much less tiring than marching along the paths of the earth" (Beauvoir (2015 [1949], p. 330).

At the *Playmode* exhibition in Lisbon, Portugal (2016–2019), and in four Brazilian cities, Belo Horizonte, Rio de Janeiro, São Paulo, and Brasilia (2019–2023),[9] co-curated by the author of this text, the aim of the artistical concept was to inquire into future possibilities for arts and play in the age of globalization. The exhibition proposes a manifesto (Lack

---

[9] https://www.tourvirtual360.com.br/ccbb-playmode/. @CCBB Brazil.

2017)[10] about interaction and participation merged with play and gaming and it shows how artists could play a role to engage people in a shared speculation about the future. Agreeing with Robinson, we can ask: "Why do we do things? What do we want? What would be fair? How can we best arrange our lives together on this planet? Our current economics has not yet answered any of these questions. But why should it? Do you ask your calculator what to do with your life? No. You have to figure that out for yourself" (Robinson 2020, p. 166).

Internet art should not be confused with art on the internet, which is a completely totally different thing. Internet art intertwines with aspects related to access to technology and cultural decentralization. Internet art creation, production and consumption belongs to a brand-new planetary world. In this way it shows, in an expressive way, how media spheres increasingly function as public space (Moss 2019). Some authors emphasize network connectivity and, therefore, adopt the term net.art instead of internet art, emphasizing network connections to the detriment of understanding the internet as a medium. However, there is already a consensus considering that internet art marks a set of works that, in their genesis, are different from previous forms of meaning creation and production. Online artefacts such as email projects, software applications for the network, individual websites or, more recently, artistic experiences and performances that take advantage of social networks, augmented or virtual reality. Thus, it became an aesthetic that radically combats the creation and production of objects and the cult of individual practices, proposing collaborative networks that value the culture of the web.

If art history and communication sciences have different understandings of a medium such as the internet this is because video art has transformed mass media devices, such as, for example, television, into an individual artistic expression medium of communication. However, internet art does not reformulate, redirect or deviate (*détournement*) only the meaning and aesthetic interpretation of the works, as another artistic tool; rather, it uses the web as an opportunity to radically change art and its processes. For communication sciences, technical objects are cultural artefacts, and the machine integrates the technological device in a broad way.

---

[10] For an updated version of the role of manifests in contemporary artistic practices please see J Lack, (Ed. 2017), *Why Are We Artists'? 100 World Art Manifestos*, UK and elsewhere: Penguin Random House Classics.

Thus, in tune with Moss, it a definition of culture is proposed that no longer allows the division between cultural and technical, as humans regulate their relationship with the world and with themselves through a process of individuation. The concrete or evolving technical object approaches a mode of existence close to natural objects, tending towards coherence and, in this sense, the technical mentality goes hand in hand with the technical culture. Ceci Moss, based on the French philosopher, Gilbert Simondon, and the Italian activist critic, Tiziana Terranova, considers that the expanded internet refers us to the concept of informational milieu and, in this context, "for Terranova, Simondon's ideas are compelling precisely because of his understanding that information is not the *content* of communication but an unfolding process within its material condition. Informational processes exist in the environment in a way that is inherently 'immersive, excessive and dynamic' and that points toward an interpretation of information that is not reduced to mere signal and noise" (Moss 2019, p. 23).

Internet arts merged global cross-cultural sensibilities with an openness to the world as a planetary and inclusive space. Unlike social media arts, internet artistic practices are, since the 1990s, based on live archives, activism, participation and the fight against the lack of inclusion on a planetary scale. Internet artistic practices can be fundamental to the integration of excluded young populations, for example, in Asia and Africa.

In 2022, two of the biggest contemporary art events in Europe, Documenta fifteen in Kassel (VV.AA. 2022) and the Venice Biennale (VV.AA. 2022), opened their strategies to the views of curators and artists from other latitudes or genders, in an economy of solidarity, attention and empathy. Documenta fifteen was curated for the first time by a collective of artists from Asia. The Ruangrupa collective[11] used the word *lumbung*, an Indonesian word that refers to a rice barn, as a concept that triggers a whole collaborative artistic practice in which a system of values is built where the accumulation of capital from crops/creations is collectively governed. Also, this year, Venice Biennale, after 127 years of history, curated by the Italian Cecilia Alemani, presented an exhibition in which women dominated the event, highlighting neglected cases while investigating themes such as gender plasticity and ambivalence, colonialism and patriarchy. The purpose of these events is to show how a more sustainable future, in terms of inclusion, is possible. The integration of those who

[11] https://ruangrupa.id/ (Accessed 27.12.2022).

have been systematically relegated to the margins in these art events is key for future inquiry.

Public life in the twentieth century, warns Ortega y Gasset (1989 [1929], p. 39), "is not only political, but, at the same time and even above all, intellectual, moral, economic, religious; it comprises all collective uses and includes the way of dressing and the way of enjoying." In this context, the crowd became visible and the masses, that is, those people who feel like everyone else and who are part of a vast group of individuals, grew up and organized themselves in movements often typical of extemporaneous humans without memory or historical awareness. Although we live today in a very different reality from that of the "classical era" of mass politics from the mid-twentieth century to the end of it. Contrary to the scenario then, when people's loyalty to traditional parties was much stronger and political battles were mainly fought over economic and state redistribution, our political systems today are facing enormous changes (Eatwell and Goodwin 2019, p. 199). American culture, according to Ortega y Gasset (1989 [1929]), had some prominence in the twentieth-century historical period by paying special attention to popular culture and artistic works from mass culture that, in a way, were catalysts of an idea of social progression making room for a world that suddenly grew and extended to the entire planet.[12] Internet arts and culture helped shape this environment and "many citizens are now living in a political world that is more volatile, fragmented and unpredictable than at any time since the birth of mass democracy. And it is unlikely that these changes will be reversed soon" (Eatwell and Goodwin 2019, p. 200). Technical objects are disguised as a commodity and, therefore, reveal their double sorcerous character, hiding

[12] If nineteenth-century civilization can be summarized, for Ortega y Gasset, in two major dimensions, liberal and technical democracy, a symbiotic copulation between capitalism and experimental science, we must also consider that not all technique is science. The Spanish author's criticism of the specialization of science is vast, even going so far as to state that science needs humans to specialize, but that it cannot itself specialize or, in other words, in politics, in art, in social uses, in other sciences, specialists can take primitive positions of great ignorance, becoming hermetic, esoteric and satisfied within their limitations and, in this sense, approaching the humans of the masses. Thus, it is concluded that specialization can suffocate both science and the state and that, for this very reason, both must be combinations of blood and languages, mestizos and plurilingualism, places of convergence where natively separate groups are forced to live together. It is therefore considered that the state should contain, above all, a program of collaboration and dynamism.

their religious reflection. Only when working conditions and practical life allow human beings to have transparent and rational relationships with their fellow human beings and with nature will the mystical cloud tend to disappear (Marx 1975 [1867]).

According to Harrison Fluss (2019), the contradiction between reason and values, or between instrumental rationality and what has intrinsic value, covers the limited forms of bourgeois reason, or positivism, which give rise to its own romantic antithesis. Within the limits of the market, one cannot see anything beyond the contradictions of capitalism.

Hegel and Marx showed that it is possible to have a dialectic of reason that can redeem what is relevant in art and religion for human life. In other words, one can extract the rational kernel of a tradition from its mystical envelope. The aim of Marx's critique was to translate religious and political struggles through "self-conscious human form." Dialectical reason, as a substantive form of rationality, can determine what human needs and interests are without superstition. But those who see reason only as instrumental will be condemned to the dilemma of looking for archaic values for their answers, instead of looking to the "poetry of the future" (as Marx said). Dialectics, as an art that rescues reason in what is apparently irrational, can help to reposition the role of popular mass culture and integrate specialists into the wider scope of knowledge, since sometimes they are effectively the ones who have the reason on their side, revealing the blindness of the elites who hold power, and so it is no wonder that Hegel and Marx saw comedy as a superior form of reason compared to tragedy (Fluss 2019). The cynic, that is, the one who has never created or done anything, parasites civilization as a "satisfied boy," using Ortega y Gasset's terminology, and bitter cynicism (Haraway 2017) proliferates in contemporary society through multiple apocalyptic visions that foster fear and inertia. Awakening the ghost of Diogenes, the one who disdains and dismantles the various power relations, perhaps makes sense (Sloterdijk 2011 [1983]).

4   Play and Gaming Can Integrate Women Studies to Generate Convergent and Sustainable Futures

The dissemination of the network culture and the internet, in the 1990s, made it possible for some women to appropriate the medium and use it to establish international connections and create works from there. The expansion of the internet has unleashed and given visibility to a set of performative practices of feminist origin, online and offline, which were made explicit through renewed and diversified types of feminisms: cyber, eco, techno, xeno... The internet without borders, boosted by globalization, has made it possible to share knowledge and create collaborative projects. Women, mediated by technological symbioses and stimulated by the need to be alert to biological and planetary sustainability, which could no longer support modern dualities, played a fundamental role in the first decade of internet implementation. Today, these practices are less dispersed and the convergence present in post-human feminism (Braidotti 2022) allows us to anticipate and predict possible futures that are based on creative and artistic speculations. With humility and determination, the arts can and should contribute to living with dignity on Earth, demystifying visions that only propose catastrophe as a possibility.

In this context, states Haraway (2017), the Anthropocene and Capitalocene histories are too stilted and both Marx and Darwin contributed to their demystification by creating possible alternatives without determinism, teleology and pre-defined plans. For Haraway, the human is a composite constituted through its relationship with life, the earth and other species, in a poetic symbiosis or *sympoiesis,* a suitable word to name situated, complex, dynamic and responsive historical systems. *Sympoiesis* involves, extends and generatively amplifies the concept of autopoiesis. If modernism was made explicit by an emphasis on the mechanism of autopoiesis, which makes living beings autonomous and characterizes them as such, the eco- and techno-feminist movements of recent decades emphasize the symbiotic relationship between multiple species as a living art, an emerging and extended synthesis that merges arts, sciences and biology. Thus, human and non-human ecologies, evolution, development, history, affections, performances, technologies and everything that is deemed pertinent are brought together. An ecology inspired by a playful feminist ethic of response-ability that promotes intricate relationships and coalitions between arts, sciences and technologies, uniting codes and algorithms, creativity, and community involvement. A proposal for artistic creation that helps us to live better on a damaged planet. The awareness that we are faced with the problem of living in a complex world suggests that human beings live and die with dignity and together on earth.[13]

The balance between reason and emotion in a biocultural continuum is shaped in contemporary feminist materialism, in reaction to postmodernism, and it rests on the premise of a nature–culture continuum that is technologically mediated through a heterogeneous ecology that includes the organic and the non-organic. Bio-power has moved to a logic of information dissemination whose bodies are transformed into techno bodies that are permeated by the vicissitudes of the environment, are socially responsible and affectively connected. These bodies are simultaneously real and virtual and are diluted through exposure and disappearance to multiple techno-biogenetic networks mediated by computing. These cyberfeminist experiments, which date back to the 1980s, were a consequence of internet and new media culture. It is important to highlight the legendary Australian art collective VNS Matrix which, in 1991, published the *Cyberfeminism Manifesto for the Twenty-first Century*. Cyberfeminism, Braidotti (2022, p. 145) points out, "is in some respects a predecessor of post-human critical feminism because it includes an intimate and productive relationship with the technological universe, breaking with the recurrent tradition that advocates a utopian hope or a deep suspicion with technology."

Techno-feminism addresses the social effects of science and technology, especially among those on the margins and the excluded, considering the eco-feminist ideal of sustainable interdependence. Studies of technoscientific feminism aims to reconcile the various souls of feminist theory with rigour and creativity. Thus, the tensions between biological essentialism and discursive constructivism are being integrated through new constellations that combine life with technology to create a new epistemology that leads to an ethics that defines new policies.

---

[13] Haraway thinks in terms of a heritage of histories and worlds situated in a concrete context in which activist practices of the arts and design integrate, in public places, diverse people and, sometimes, other creatures and species. For most species and communities, play was the most powerful and diverse activity for reorganizing old things and proposing new ones. Thus, new patterns, feelings and actions for creativity emerge that can be triggered safely and/or through conflict and collaboration. Haraway also suggests that we nurture a vital memory and that we don't forget the foul smell in the air from the burning of witches, that we don't forget the murder of human and non-human beings in the Great Catastrophes of the Plantationocene, Anthropocene and Capitalocene. For that purpose, we should mourn the dismemberment of the world and we may have to learn how to talk to the dead.

For Moss, today's internet, compared to that of the 1990s, is more mobile, ubiquitous and mainstream. Artists and critics react to these changes. Moss suggests, "recognizing the decline of Clement Greenberg's concept of medium specificity that understood art according to the essence of its medium as determined by its material properties, and the difficulty hybrid media (like the computer) present to the establishment of 'pure' art forms like painting or sculpture, Krauss argued that instead artists must 'purify' art itself away from the infringement of fashion and kitsch" (Moss 2019, p. 25). Answering the question "what can art be in the context of the 21st century?," Ceci Moss puts forward the idea of an expanded internet art, something like a continuous that exists in a distributed system, an artistic work without object or centre, without an autonomous and singular existence. In that fashion, internet art is always in motion, always in circulation, in assembly and dispersion (ibid., pp. 38–39). The concept of flux describes immanence as opposed to transcendence, where the milieu is a condition of existence. Thus, the living organism participates in and helps to shape the milieu.

If cybernetic theories had mechanistic tendencies, their application to human societies, through systems theory, extended the idea or belief that the universe works in the manner of automatons, forgetting that science itself maintains an intrinsic relationship with perception. The immanent conditions of individuation are indeterminate and reside in a process of invention that is not predictable, but rather is open to disturbances, improvisations and adjustments in a generative chaos. Information does not exist by itself without the meaning generated by individuals and any technological invention emerges in tune with the needs and experiences of humanity. Through a process of ontogeny, there is a symbiosis between technology and the contemporary informational environment. The informational milieu restructures the production and existence of creative expression, and a paradigm shift seems to arise where meaning is no longer about signs, but rather about milieu signs. The process of permanent innovation presupposes a divorce between the rhythm of cultural evolution and the rhythm of technical evolution since technique evolves more than culture. Anchored in Bernard Stiegler, Moss suggests that we should stop considering technology as purely utilitarian or a threat to humans and that we must focus instead on thinking about how the explosion of new technologies alters the human experience.

5 Conclusion: Speculative Arts-Based Research: From Objects to Processes, Resistance Against Modern Delusions

Contemporary artists work productively from the symbiosis between information technologies, the environment or context in which they are inserted and human experience. Moss's analysis of *Les immateriaux* exhibition, by Jean-François Lyotard and Thierry Chaput, which took place in 1985 at the Center Georges Pompidou in Paris, should be highlighted. The concept of this exhibition takes us back to Jean-François Lyotard's concern with the power of new technologies to inscribe themselves in collective human memory, in a time and space in which this inscription will promote deterritorialization. Lyotard worked to promote the idea of an exhibition as a work of art, showing how connections between matter, mind, time and memory are reconfigured, and highlighting how new technologies extend our capacity for memory while working on a scale outside the human perception. At the *Playmode* exhibition in Portugal and Brazil, our aim was to show how central play, gaming, interaction and participation were key concepts in shaping human contemporary perception and enhance the ability to resist.

Resistance is an art that bears witness without prejudice; it is generated not in reaction but rather in relation to contemporary circumstances and developing sciences and technologies. Without immediate judgments or decontextualized epistemological biases, it is intended to generate greater attention and care through an approach that considers the fluidity, performance and context of identity categories, emphasizing social contextualization instead of social categorization (Shaw 2014), since the latter is based on an old-fashioned idea of identity as an essence so widely criticized by postmodernism. We are dealing with what Ndikung called performative knowledge events that promote spaces of care. These spaces emphasize relationships and multiple perspectives between things, people, thoughts and forms of knowledge, bringing together areas such as curatorship and anthropology, which end up being more than the sum of the parts that make up the relationship. The illusions and disappointments of care and empathy suggest spaces for sharing and exchange in constant evolution and mutation and are no longer limited to the cult of artists and works of art that, after all, do nothing more than feed the failing neoliberal market of contemporary arts (Ndikung 2021).

For Fernando Domínguez Rubio (2022), the unnatural ecologies of the modern museum operated through the primacy of the artistic object which, according to this dogma, must be a readable, singular and authentic product of the artist's creative intention and agency. Now, digital objects put this paradigm in crisis through displacements that denounce their fragility as originals and, often, as collective works, promoting, as an alternative, notions of copies always predisposed to updating to avoid obsolescence. These aspects cause problems and challenges to conservation, as digital objects operate in a different regime from that of modern artistic aesthetics, refusing the separation between subject and object of enjoyment. In this context, a type of interpretive conservation will be necessary, which leads us to the ability to interpret multiple iterations of the same work, as already happened, for example, in the history of games and performance. Thus, it is possible that the work is kept through various living and documentary records that refer us to different perspectives to preserve that memory in a hybrid way. This memory (record) can consist of images, texts, movements, interactions, sounds… The idea of a stable object gives rise to the artistic work in constant circulation, which multiplies and regenerates in channels of equivalent objects. To survive, the work will then have to continually regenerate itself, move from a discrete object to a distributed object and, for this to happen, conservators, curators, technicians and artists are called upon to interact with each other to give shape to projects each time more complex. We are witnessing, through digital culture, a paradigm shift where what was considered modern artistic aesthetics, that is, the unique, singular and authentic object, is no longer applicable and, therefore, we are faced with a different artistic order that is based on circulation, dissemination, and constant creation of partial and fluid objects.

In this dynamic environment we should realize that reason and emotions are not separate entities and we should maybe acknowledge the power of stories in shaping cultural values and political societies. As Ursula K. L. Guin work poetically demonstrated, happiness is related to reason and only by reason can it be conquered (Le Guin 2018 [1969], p. 221). Because (ibid., p. 216),

> Light is the left hand of darkness
> And darkness, the right hand of light
> (…).

REFERENCES

Ascott, R. 1990, Is There Love in the Telematic Embrace? Art Journal Vol. 49, No. 3, Computers and Art: Issues of Content (Autumn, 1990), pp. 241–247 (7 pages) Published By: CAA. Retrieved July 21, 2022, from, https://doi.org/10.2307/777114

Beauvoir, S. de 2015 [1949], *The Second Sex*. Lisboa: Edições Quetzal, Serpente Emplumada.

Bolter, J. D. 2019. *The Digital Plenitude: The Decline of Elite Culture and the Rise of New Media*, Cambridge: MIT Press.

Braidotti, R. 2022. *Posthuman Feminism*. Cambridge: Polity Press.

De Landa, M. 2000. *A Thousand Years of Nonlinear History*. New York: Swerve Editions.

Eagleman, D. 2021. *O Cérebro em Ação, Nos bastidores do cérebro em constante mudança*. Lisboa: Edições Lua de Papel.

Eatwell, R. & Goodwin, M. (2019), *Populismo, A Revolta Contra a Democracia Liberal*. Porto Salvo: Edições Saída de Emergência.

Ellis, C., Adams T. E., & Bochner, A. P. 2011. Autoethnography: An Overview. Forum Qualitative Social Research. Volume 12, No. 1, Art. 10 – January 2011. Retrieved July 21, 2022, from, https://www.qualitative-research.net/index.php/fqs/article/view/1589/3095

Fluss, H., 2019. The Nightmare Before Socialism, Jacobin Magazine. Retrieved July 22, 2022, from, https://jacobinmag.com/2019/10/nightmare-before-christmas-halloween-socialism

Gouveia, P. 2019. Play and games for a resistance culture [Brincadeiras e jogos para uma cultura da resistência], *Playmode* Exhibition Publication, Fundação EDP, MAAT Museum, English and Portuguese, pp. 8-27.

Gouveia, P. 2022. The strange world of *Playmode* Brazil: playful sounds and complex critical speculations [O estranho mundo da *Playmode* Brasil: sonoridades lúdicas e especulações críticas complexas], Playmode Brazil Publication, Brazil Bank Cultural Centres (CCBB), English and Portuguese, pp. 17–33.

Haraway, D. 2017. *Staying with the Trouble*. London: Duke University Press.

Kelly, K. 1995 [1992]. *Out of Control, The New Biology of Machines*. London: Fourth State.

Lack, J., 2017. *Why Are We Artists? 100 World Art Manifestos*. London: Penguin Random House Classics.

Le Guin, U. K., 2018 [1969]. *A Mão Esquerda das Trevas*. Lisboa: Relógio D'Água.

Lees, A. J. 2022. *Brainspotting, Adventures in Neurology*, Notting Hill Editions, Britain, eBook, Machine Learning, 5-6.

Marx, K. 1975 [1867]. *O Capital*. Viseu: Edições Delfos.

McLuhan, M. & Powers, B. 1992. *The Global Village*. Oxford: Oxford University Press.

Morris, I. 2010. *Why the West Rules—For Now, The Patterns of History, and What They Reveal About the Future*. New York: Farrar, Straus and Giroux.

Morton, T. 2018. *Dark Ecology, For a Logic of Future Coexistence*. New York: Colombia University Press.

Moss, C., 2019. *Expanded Internet Art: Twenty-First-Century Artistic Practice and the Informational Milieu* (International Texts in Critical Media Aesthetics), Bloomsbury Publishing Academic: New York, London, Oxford, New Delhi, Sydney.

Ndikung, B. S. B., 2021. *The Delusions of Care*, Berlin: Archive Books.

Nicolelis, M. 2021. *O Verdadeiro Criador de Tudo*. Amadora: Edições 20/20 – Elsinore.

Nora, S. & Minc, A., (1978), *L'informatisation de la société: rapport à M. le Président de la République*, Volume 1, Éditions La Documentation Française. ISSN 0554-5412.

Page, M. 2002. *The First Global Village. How Portugal Changed the World*. Lisboa: Editorial Notícias.

Preciado, P. 2019. *Manifesto Contra-Sexual*. Lisboa: Orpheu Negro.

Robinson, K. S. 2020. *The Ministry for the Future*, Orbit: Great Britain.

Rubio, F. D., (2022), Keynote speech: The Unnatural Ecologies of Modern Art, online symposium: Computer legacies: Narrating histories of digital media in museums, Loughborough University

Schrage, M. 1995. Revolutionary Evolutionist, For Richard Dawkins, genes are selfish, the watchmaker is blind, and the mystery of life is no mystery – it's digital. Retrieved January 20, 2022, from, https://www.wired.com/1995/07/dawkins/

Shaw, A., 2014. *Gaming at the Edge, Sexuality and Gender at the Margins of Gamer Culture*. Minneapolis & London: University of Minnesota Press.

Sloterdijk, P. 2011. *Crítica da Razão Cínica*. Lisboa: Relógio D'Água Editores.

Turkle, S. 1997. *A Vida no Ecrã, a identidade na era da internet*. Lisboa: Relógio de Água Editores.

Turkle, S. 2021. *The Empathy Diaries*. New York: Penguin Press.

VV.AA. 1997. *Catálogo Jovens Criadores'97*, Lisboa: Edições Clube Português de Artes e Ideias.

VV.AA. 2022. *Documenta Fifteen Handbook*, Hatje Cantz, Kassel.

Woolf. V. 1947. *Three Guineas*. London: Hogarth Press.

Yong, E. 2021. Neuroscientists Have Discovered a Phenomenon That They Can't Explain. In *The Atlantic* Retrieved July 21, 2022, from, https://www.theatlantic.com/science/archive/2021/06/the-brain-isnt-supposed-to-change-this-much/619145/

The Mind, Embodiment and Technology: Experience, Aesthetic Sensibility and Production of Meaning

Ralph Ings Bannell

1   INTRODUCTION

One of the hottest topics of current debate are the so-called intelligent technologies that include, amongst others, Artificial Intelligence (AI) image generators, large language models, machine learning and natural language processing. Apart from the development and application of these technologies in almost all dimensions of life, there is a heated debate in philosophy and cognitive science about the possibility of general AI, that is, the possibility of such artificial systems becoming conscious and acquiring the same cognitive capacities as human beings.

Some think this possibility is not far away (Kurzweil **2005**), while others think it is on the very distant horizon but still theoretically possible. In this chapter I will argue that such predictions cannot be realized unless the artificial systems have organic bodies and aesthetic sensibilities, which

R. I. Bannell (✉)
Catholic University of Rio de Janeiro, Rio de Janeiro, Brazil

© The Author(s), under exclusive license to Springer Nature Switzerland AG 2024
P. Alexandre e Castro (ed.), *Challenges of the Technological Mind*,
New Directions in Philosophy and Cognitive Science,
https://doi.org/10.1007/978-3-031-55333-2_10

make it possible to have experience and produce meaning in ways similar to human beings.

My argument will be that artificial systems and so-called "intelligent" technologies cannot produce meaning or have experiences because they are not embodied systems. After looking briefly at the dominant present-day model of mind, I will look at the concepts of experience, aesthetic sensibility and production of meaning through the lens of John Dewey's work and the work of the philosopher Mark Johnson, influenced by his thought. Hopefully, this analysis will show why artificial systems and "intelligent" technologies are not capable of aesthetic sensibility, cannot be producers of meaning and, therefore, subjects of experience.

2   The Dominant Model

The predominant paradigm of the human mind is based on a series of dichotomies, especially subject–object, mind–body, mind–nature, cognition–emotion, individual–collective, internal–external, cognition–action and theory–practice, among others. The central idea, with its roots in early modern philosophy, is that the human being is a disengaged subject (Taylor 1989), with his or her mind being separated from the environment (physical and social), body and nature. A consequence of this conception is to separate what is supposedly internal to the subject from what is external to the organism. Thus, what is important for understanding the human mind is located internally, the location nowadays being the brain, with the environment seen only as a source of stimuli from sensory modalities and the body only as a vehicle to carry these stimuli to a supposed internal cognitive mechanism.

These internal mechanisms are usually understood as computational and involve the processing of information that comes from the outside, so to speak. This implies that the information received through stimuli already has meaning, which then has to be decoded and reconstructed by the organism's brain. Of course, it is possible to keep the computational model and include external factors as constitutive of mind. For a classic view of this kind, see Hutchins (1995). Clark (2003, 2008, 2014) tends in the same direction with the extended mind hypothesis, which I will mention later in this chapter. Even when the model of linear computational processes operating over symbols is rejected, for example in parallel processing models and connectionist neural networks, the assumption is that some kind of computational process is occurring and that this process can be simulated in artificial systems. It is not difficult to see why these assumptions can be found in artificial models of mind. These models think of cognition as the internal processing of information acquired from the environment as stimuli, which are then decoded and reconstructed by internal rules or algorithms that operate on an inductive logic of probability. By analogy, the human mind is understood as operating on the same logic and model.

Without trying to discuss all of these themes and philosophical aspects, something which is impossible in a single chapter, I would like to elaborate a discussion of three central concepts: experience, aesthetic sensitivity and production of meaning. Experience, because it is a central concept for theories of mind and needs to go beyond the notion of the stimulation of sensory modalities; aesthetic sensitivity, to reinstall the body, feelings and emotions in the analysis of the mind and human cognition; and production of meaning, to free this term from the exclusive connotation of a process of information processing, which does not involve the body in movement, the materiality of the environment or affectivity.

3   A Starting Point for Thinking About Experience

In their book *Radicalizing Enactivism*, Hutto and Myin (2013, p. 15) contest the traditional conception of experience, saying that "perception cannot be accounted for solely in terms of raw stimulation and perturbation of sensory modalities." They distinguish between sensing and experiencing, defining 'sensing' as simply receiving stimuli through the sensory organs and 'experiencing' as something that requires elements beyond such stimuli. The term 'experience' conveys the idea of a "full-blown experience of worldly offerings," which is the phenomenon that Hutto and Myin try to analyze in their book.

These authors argue against the idea that experience, at least basic experience of the world such as perception, has semantic content or, indeed, any representational content. This thesis is controversial, of course, but it is important to note that the target of its attack is a well-defined conception of semantic content, that is, a mental representation that satisfies truth conditions or, at least, adequacy conditions. I agree with the rejection of this notion of content as necessary to explain all perceptual experience. I also agree with the rejection of concepts and propositions as essential to such experience. If it is possible that perceptual experience, for example, has no content of any kind, as Hutto and Myin argue, that is another issue, which I will not discuss here.

I would like to suggest that experience does not need to have propositional-conceptual content to have meaning for a perceiver responding to the world. In other words, I think we can accept, with Hutto and Myin, that experience does not always need to attribute semantic properties to objects and situations in the world, in the sense of having truth or satisfaction/precision conditions, being a propositional attitude or even containing concepts. However, experience is still the production of meaning, even without these conditions. I see no problem with referring to this meaning as its content and referring to the process that produces it as the production of meaning. One author (Gumbrecht 2003) calls this process "the production of presence" instead of meaning, precisely to avoid confusion with traditional semantic theories, including those in the interpretative tradition of hermeneutics. I prefer to stick with the term 'production of meaning', with the appropriate caveats elaborated above.

Hutto and Myin (2013, p. 81, my emphasis) advance the proposal that "organisms often act successfully by making appropriate responses to objects or states of affairs in ways that are *only mediated by their sensitive response to natural signals*, where this responding does not involve contentfully representing the objects or states of affairs in question." The authors also state, "that things look and feel a certain way does not entail that perceptual states possess or attribute content. Perceiving (…) lacks inherent conditions of satisfaction" (Hutto and Myin 2013, p. 134). I agree that content, defined semantically in the way Hutto and Myin define it, does not need to be present in the production of meaning by an organism coupled with its environment. If this is so, how can we explain the phenomenon of perceptual experience? Hutto and Myin do not discuss the aesthetic dimension in their analysis of experience, at least not in the book mentioned here. That's what I intend to do in this essay.

I would like to present an analysis of the basic experience and production of meaning that starts from aesthetic sensibility, therefore offering an extension of the explanation of experience that does not involve representations or semantic content in Hutto and Myin's sense. My analysis will be a form of Enactivism, therefore, in the same line as Hutto and Myin, but in dialogue especially with American Pragmatism and Autopoietic Enactivism, strands of thought either ignored by Hutto and Myin (in the case of the first) or to which they give little attention (in the case of the second).

4   Experience

I begin with the concept of experience in Mark Johnson, a philosopher who has developed incisive analyses based on the insights of John Dewey's Pragmatism. Dewey's concept of experience is notoriously difficult and even vague, despite its centrality to this philosopher's writings. Johnson helps develop and clarify this central concept, developing analyses in several books over the last few decades (Johnson 1987, 2007, 2017, 2018; Johnson and Tucker 2021; Johnson and Schulkin 2023) in dialogue with contemporary cognitive sciences. The result is a conception of experience that involves the brain, the body in movement and the environment in its analysis. For Johnson:

> Experience is what happens when an active complex organism engages its multidimensional environments. Experience is neither exclusively subjective nor objective, cognitive nor emotive, theoretical nor practical, mental nor physical. (…) Experience is all these dimensions interwoven—not as ontological or epistemological dichotomies, but as inseparable yet distinguishable threads of an ongoing process of organism-environment interactions or transactions. Experience thus includes both *what* is experienced as well as *how* it is experienced (…) and it encompasses every aspect of our bodily being in the world. (Johnson 2018, p. 207, emphasis in the original)

Immediately, we see the attempt to overcome the dichotomies mentioned at the beginning of this chapter. An experience is a quality, and how it is experienced is through aesthetic sensibility by the organism, in ways that do not necessarily involve verbal language, concepts or even symbolic signs. To understand this, we need to go back to Dewey and his analysis of how we produce meaning in interaction with the world. Dewey argues that advanced cognitive abilities, such as thinking, are rooted in more primitive experiences that we have when we engage with our environment. At the beginning of his article "Qualitative Thought," Dewey (1930/1984, p. 243) says:

> the world in which we immediately live, that in which we strive, succeed and are defeated is preeminently a qualitative world. What we act for, suffer and enjoy are things in their qualitative determinations. This world forms the field of characteristic modes of thinking, characteristic in that thought is definitely regulated by qualitative considerations.

But what are the "qualities" that permeate the environment for Dewey and regulate thought? Unfortunately, Dewey is not as clear as he could be about the qualities we encounter when we interact with the world. At one point he says that these qualities are not something exclusively internal to the feeling subject, but are qualities in which things and organisms participate, making it impossible to separate the qualities from the organism that feels them (Dewey 1930/1984). In other words, qualities are not only in the environment or only in the subject; they are relational and enacted in the coupling of the organism with its environment.

But are qualities simply particulars we encounter in the world? This raises the normative question of how we are guided by qualities that already exist, so to speak and, indeed, how they can exist independently of individuals. Can these qualities be understood as patterns that regulate our interaction with the world? At the same time as Dewey, but independently, Merleau-Ponty also said that "each organism, in the presence of a given milieu, has its optimal conditions of activity and its own way of achieving balance," and each organism "modifies its milieu according to the internal norms of its activity" (Merleau-Ponty 1969, pp. 148, 154).

There is obviously a need for norms here, for perception and action, otherwise we would have to enact qualities based on singular encounters with particulars in the world, something that would make even the simplest of acts almost impossible. According to Dewey, to the extent that certain objects and events, etc. become important for the survival of human beings—in the prevailing situations in which they need to live—these objects and events are fixed and identified by the sensorimotor patterns of movement and in the interaction between the organism and the environment, being sedimented as established norms in the constant interaction between organism and environment, and subsequently guide a particular individual's action. This applies both to the meaning created by feelings and sense and to the meaning created by the verbal language used between individuals, in the formulation of concepts, language and communication.

Dewey (1930/1984) gives us some insight into how we make sense of the world and survive in it by discriminating certain objects and their properties from a situation we confront. We do this, firstly, through feelings, resulting from the aesthetic sensibility we have towards the "felt qualities" of the situations in which we find ourselves.

The feelings Dewey refers to "blend with the general situation"; they are "tones of the overall tone of a situation," in the background, so to speak (1926/1981). Another dimension of meaning Dewey called "sense," a form of meaning that has a reference, but that cannot yet be formulated into a proposition, because we have not yet discriminated properties and relations that could be articulated with concepts and included in propositions. Dewey reserves the term 'signification' for the meaning created by verbal language, developed in the evolution of human beings because of the complex interactions we have with the environment and which require a more complex regulation of our interaction with the world. However, the other terms he uses ('feeling' and 'sense') for our capacity to interact with and comprehend the "felt qualities" of a situation are extremely important for our understanding of how we make sense of the world. This is because they are grounded in our bodily and affective capacity to react to the world.

It is important to emphasize that there is no metaphysical necessity operating here. Established patterns of meaning, in all of the connotations mentioned here, are contingent. The fact that we discriminate between objects of different types is a matter of the meaning attributed to those objects by human beings who have a certain sensitivity to the environment and need to pay attention to objects and events for their survival. Both sense-making and its normative dimension emerge from the coupling of the organism with its environment, a precarious situation within which the organism has to choose and create the necessary conditions to achieve equilibrium and adapt to the environment through its own agency. Of course, it is not the case that each individual does this *ex nihilo*. The infant is born and grows up in an environment where standards and norms are already established by previous generations and in which his or her organic constitution makes possible a sensitive interaction with the world. Of course, these standards and norms can be criticized and/or modified. But, to the extent that they are sedimented, they act as (partial) determinations of our experience of the world. The important point here is that we are organisms for which things matter. In contrast, nothing matters to an artificial system. I'll come back to this point later.

One consequence of this is that certain experiences are historical and culturally sedimented. They capture the felt qualities of a time or culture. We see this clearly in paintings, an analogy also used by Dewey. Vermeer's works, for example, capture the felt qualities of everyday life in seventeenth-century Holland. Picasso's works do the same thing for the felt qualities of the twentieth century. Our reaction to these works depends on our aesthetic sensitivity. It is more difficult to have an aesthetic reaction to a work of art from a previous epoch because we no longer inhabit the environment in which it was produced. Therefore, its meaning is not clearly captured by our sensibilities. The same thing can happen with works from different cultures, when the sensitivity to capture their meaning is not available to an individual from another culture. As much as our world is (partly) historically constituted, it is not homogeneous. Watching a film from another culture, for example, can produce perplexity, which is sometimes felt as discomfort, for someone who does not share that culture.

It is our aesthetic sensibility that, according to Dewey, allows us to perceive the qualities that permeate any situation and, therefore, perceive the world and attribute meaning to it. For Dewey, this sensitivity is then transformed into feelings, a kind of susceptibility to the useful and harmful aspects of our environment. It is its manifestation as feelings and emotions that produces the valences that lead to action.

And, as counterintuitive as it may seem, our aesthetic sensitivity, our feelings and our emotions guide our reasoning too, even the most sophisticated and abstract. Furthermore, both aesthetic sensitivity and emotions and their normative dimensions change historically, creating new norms for the production and understanding of meaning. What we need to understand for now is that it is through our aesthetic sensibility, expressed in our feelings, and our affective response to the world, during our bodily engagement with it, that we are able to make sense, on a basic level, before we use verbal language to express and communicate concepts and thought.

5   Aesthetic Sensibility

Starting once again from Mark Johnson's (2018, p. 7) analysis, we can say that "aesthetics is fundamentally about how we are able to have meaningful experience." He goes further, saying that "we have a deep visceral, emotional and qualitative relationship with our world (…). This visceral engagement with meaning (…) is the proper scope of aesthetics" (Johnson 2018, p. 1). This focus on the visceral and the aesthetic brings the body to the center of the analysis of experience.

It also brings the body to the center of the analysis of meaning production. Once again, Johnson, following Dewey, argues that meaning is based on bodily experience. It arises from the qualities we feel and the sensory patterns our bodies acquire over time. Of course, it is not limited to bodily engagements because we produce meaning with language as well (perhaps the most sophisticated kind of meaning) but our bodily engagement with the world is always the ground of meaning; meaning depends upon the qualities of situations and our bodily experience of them.

This is important because Analytical Philosophy, especially, has reinforced the mistaken idea that all meaning is a matter of verbal language and that only words and phrases have meaning. It is also assumed that only sentences have meaning because they express propositions using concepts, which map states of affairs in the world. In short, meaning is seen as a matter of how words and sentences can represent the world (have reference and meaning); that is, meaning is seen as propositional-conceptual and involving truth conditions (Johnson 2018, p. 243). We need a concept of 'meaning' that goes beyond this model, as Johnson himself argues. If we do not accept the restriction of meaning to this mistaken model, and to semantics in the sense of linguistic meaning, we open the possibility of finding it in experiences that do not involve verbal language, nor conceptualization, in their constitution. This is exactly what Johnson does in his work.

For Johnson, then, meaning emerges from our embodied experience, especially its structure, qualities and felt direction. This means that meaning is first linked to our affective, sensorimotor processes, which have valence and an emotional charge. The meaning of the world for us is, in the first instance, produced by sensorimotor and affective processes and other forms of meaning are extensions of these primordial meanings. Johnson speaks of "schematic images," which are the way in which the organism registers the meaning of something in the world. An example of an image-scheme would be "inside-outside," produced by the repeated movements of putting things inside containers and then taking them out again. This is how concrete concepts are formed in a child's mind, for example. Afterwards, such concrete concepts are expanded, through mechanisms that include metaphorical extension, to abstract concepts. For example, the concepts 'inside' and 'outside' will be expanded to the abstract concept of 'containment.' Furthermore, metaphors are considered basic cognitive elements, without which it would not be possible to talk about the world. For example, the concept of containment is expressed metaphorically in the phrase "he is trapped by a certain idea."

The main point here is that it is embodied experience that is the primary source of meaning. It is not the sensory stimulus, which supposedly carries with it a meaning (or information encoded in some way) of something in the world that is then reconstructed by the organism's brain, as if it were meaningful information to be processed by an internal intellectual machine. Rather, basic meaning is constituted in the relationship and coupling of the organism with its physical and social environment. I think Johnson's explanation helps fill in the details of Hutto and Myin's (2013, p. 81) suggestion when they say "organisms often act successfully by making appropriate responses to objects or states of affairs in ways that are only mediated by their sensitive responding to natural signs, where this responding does not involve contentfully representing the objects or states of affairs in question." This sensitive response can be understood as the production of embodied meaning, as analyzed by Johnson.

6   COUPLING, AFFORDANCES AND MEANING PRODUCTION

Of course, for meaning to be produced in this way the organism needs to be coupled to the environment and such coupling is not arbitrary. Appealing to James Gibson's theory of affordances (1979/2015), we can say that the environment 'affords' certain couplings and not others. Referring to a concept from Uexküll (2020), we can say that the organism builds an *umwelt* from its couplings with the world, a niche in which it is easier to survive. Thus, the meaning of an object is based on the affordances for possible experiences related to that object. Furthermore, our descriptions and analyses of affordances need to contain the crucial role of the qualitative dimension of any experience, which is its aesthetic dimension.

Here we are reminded of Dewey's analysis when he starts from the concept of situation. Unlike most thinkers before and after him, he does not start from already discriminated objects, with referents, properties and relationships, and then explain an experience as the construction or assimilation (synthesis) of these atomic parts to produce something more complex, be it a concept, proposition or other meaning-carrying entity. Unlike this, Dewey starts with an undifferentiated situation and explains the process of producing meaning as one of discriminating the elements contained in the situation. In other words, the process is the opposite of synthesis or association, which is assumed as the predominant mechanism by many philosophers.

Instead, Dewey says that we always start from a situation that has qualities that permeate it and give the situation its overall quality. As already said, an analogy here could be a painting, which always carries qualities that permeate it, to the point where we generally recognize the artist from the qualities that permeate his or her works. In a somewhat similar vein,

Taylor (2016) speaks of a work of art "portraying" something. Our experience of the world is "portrayed" in the work in a way that produces a feeling in us, as we saw in Dewey. Dewey goes further, saying that our feelings vary, depending upon the directions and phases of an activity. Activities can be started or carried out in various manners and they can be gratifying or frustrating, depending upon how we are connected to the environment. Consequently, our feelings can vary as the activity proceeds.

In addition, and in reaction against the analysis accepted at his time (and for a long time after), that experience is the result of processing information received by sensory stimuli, Dewey (1896, pp. 358–359. Emphasis mine) says:

> Upon analysis, we find that we begin not with a sensory stimulus. But with a sensorimotor coordination, the optical-ocular, and that in a certain sense it is the movement which is primary, and the sensation which is secondary, the movement of body, head and eye muscles determining the quality of what is experienced. In other words, the real beginning is with the act of seeing; it is looking, and not a sensation of light.

This idea points in the direction of analyzing how the body and the environment together constitute certain cognitive capacities, in this case perception of the world. It also points in the direction of understanding meaning production as co-constituted by the organism and the environment, a topic I will now turn to.

7   Embodied Schemas

Johnson uses the term "embodied schemas" to name the bodily structures necessary for sensorimotor coordination. They are patterns of recurring structures of experience that humans (and other animals) have through bodily engagement with the environment. In other words, the contours of feelings are embodied as sensorimotor schemes in the body of the organism that reacts to the world with affectivity.

The term "affectivity" deserves some attention here, because it can refer not only to the feelings and emotions generated in engagement with the world. As Colombetti (2017) reminds us, the term means anything and everything that affects the organism in its coupling with the environment. As she herself says:

"affectivity" refers to the capacity or possibility of having something done to one, of being struck or influenced (the term comes from the past participle of the Latin verb *afficio*, "to strike, to influence"—itself a compound of *ad*, "to", and *facio*, "to do"). This influence is not merely physical or mechanical (as when one says that the daily amount of sunlight affects the air temperature) but psychological. It refers to the capacity to be personally affected, to be "touched" in a meaningful way by what is affecting one. In this broad sense, it is not necessary to be in a specific emotion or mood to be in an affective state; one is affected when something merely strikes one as meaningful, relevant, or salient. (Colombetti 2017, 448)

In other words, things strike us as meaningful in our encounters with the world. Without this dimension, nothing would have any meaning; it would be mere sensory stimulus without meaning.

Can we safely say that artificial systems are not affected in this way by anything in the world, that nothing matters to them? Is this an important aspect of human intelligence and cognition that cannot be simulated in artificial intelligence? Colombetti's definition of affectivity could be interpreted by some as including artificial systems in its scope. When a neural network "reacts" to inputs and alters itself (as in machine learning systems, for example), can we attribute affectivity to it, in this broad sense? But note that Colombetti says that the influence here is not mechanical or merely physical but psychological. And it is implausible, at the least, to attribute psychological reactions to inanimate structures. I would suggest that for something to strike one as meaningful, it has to be created by an aesthetic sensibility that machines do not have. Indeed, Colombetti (2017, p. 2 my emphasis) states that "enactivism holds that all *living organisms*, as autonomous and adaptive, are sense-making systems and that, as such, they are cognitive." Cognition as sense-making is restricted to living organisms and not artificial systems.

What we need to understand for now is that it is through our aesthetic sensibility, expressed in our feelings and our emotions, during our bodily engagement with the world, that we are able to make sense, on a basic level, before we even use concepts and verbal language to express and communicate thought. The aesthetic dimension of understanding, then, is present since our most primitive experience of the world. It is in the coupling with the world that we feel it and this feeling-the-world is a primordial way of making sense. Johnson (2007) goes further when he argues that emotions are our main means for us to be in contact with the world.

They make aspects of the environment present to us and, therefore, meaningful in the most basic sense of this term, because our bodily state shapes the contours of our experience. All of this takes place on a pre-reflective level but, nevertheless, has profound meaning for us.

8   Embodied Schemas and the Production of Meaning

If what has been said so far is on the right track, we can begin to understand how embodied schemas, the most primitive and primordial meanings, can help in our attempts to understand the world and survive in it. What is the relationship between embodied schemas and our understanding of the world?

Johnson (2018) elaborates on some embodied schemas that are central to our interaction with the world. First, he urges that we start with movement. As he says: "we learn the contours of our world and the possible ways of interacting with it through movement." Maxine Sheets-Johnstone (2011, p. 114) says something similar when she states That "movement is the generative source of our primal sense of aliveness and of our primal capacity for sense-making." As she says, we are in movement since before birth, not just the disposition to move but the real thing.

If we start with movement, we can see how embodied schemas, created in movement, can lead us to interact with the world in ways that allow for successful engagement. For example, balance is a body schema that refers to the body's internal homeostasis and physical balance. If we remember Gibson's concept of 'affordances,' mentioned earlier, we can say that physical structures in the environment 'afford' certain movements and not others; they afford balance, for example, and, when they do not afford balance, we get disequilibrium. Phylogenetically and ontogenetically, with repeated movements within physical environments, human beings develop a sense of bodily balance. As Johnson (2018, p. 218) says: "given our upright stance within a gravitational field and our proprioceptive and kinesthetic senses, we humans have developed a sense of bodily balance as key to successful transactions with our world." Such transactions require spatial orientations, for example, verticality, and the sense that the world is (at least most of the time) up and down.

These body schemas and the spatial orientations they promote allow for certain experiences within space. Johnson gives the example of architecture. Entering a cathedral, for example, with its vaulted ceiling, we not only orient ourselves by the columns going towards the sky, but we can also produce meanings of something beyond our human proportions, that is, a "spiritual place," through the metaphorical extension of the notion of balance. We often say, for example, that entering such places is an "uplifting experience." It is no surprise that cathedrals are built in this way, to stimulate an experience of looking beyond the world in those who enter them.

The cathedral is an example of how our physiological structure, together with the structures of space, can produce meanings and experiences that would be impossible without these characteristics of the body and the environment. In the same way, we can understand why some interactions are impossible for us. For example, before inventing the airplane, it was impossible for humans to have a bird's-eye view of the world. It is still impossible without machines that help with this. For the same reason, it would not be possible for a human being to live in the environment imagined by Escher in his lithograph "Relativity," shown in Fig. 10.1.

The sense of balance developed in the interaction between organism and mind cannot be used to guide us in this space, except on the plane closest to the viewer (where the figure is starting to climb the stairs). If this world existed, it would be impossible for human beings to be successful in

Fig. 10.1   "Relativity" by Escher Source: Wikipedia

their transactions with it. As Johnson (2018, p. 255) puts it: "Every architectural structure will present us with a felt qualitative unity of the whole that, in essence gives us a world (however small), and a certain way of inhabiting that world. (…) The building's particular affordances—based on its particular structures, forms and qualities—provide the possibilities for meaningful engagement with the building or constructed space."

One more embodied schema analyzed by Johnson needs to be mentioned here, which is the affective contours or patterns of feeling that we create during the process of human development. Once again, Johnson (2018, p. 215) is insightful when he says the following: "Emotions arise within the flow of our ongoing experience with our environment and are a primary way by which we assess the quality, meaning, and development of our experience. (…) [E]motions are a key part of how we gauge the meaning of what is happening to us."

We saw in Dewey the importance of feelings and emotions for the production of meaning. Our emotional reaction to the world is the most basic way of evaluating our experience as well as a contour of the experience itself. Taylor (2016) also says something similar when he states that what he calls "intuitive reactions" are ways of producing "human meanings" as well as evaluating these same meanings. I will come back to this.

9   Art, Technology and the Production of Meaning

How should we think about art and technology in light of the discussion above? In a similar way to Dewey (1934), Johnson (2018) argues that a work of art is grounded in our everyday aesthetic experience when he says: "The performance of the work of art has meaning in the same way, and through the same neural and embodied means, as our 'ordinary' experience is significant." Works of art can provide us with possible ways of being and inhabiting the world and, by extension, mold our everyday existence. A good work of art opens up possibilities for growth and enriches meaning rather than simply reproducing meanings that already exist.

Music is a good example of what is at stake here. When we listen to music, patterns of feeling are directly experienced as a "tonal analogue of emotional life" (Langer, apud Johnson 2018). Yet again: "[W]e are drawn in and carried along by the music, not just in intellectual comprehension, but also through our whole animate bodily feeling of the affect contours enacted in the music. We are swallowed up in the music, moving and feeling in and through the music. Music can give us an experience, on steroids, of our vitality and the felt contours of our emotional being" (Johnson 2018, p. 220).

How exactly this happens is still a mystery as far as I know, but the phenomenological experience is something that every human being feels. Music is a way of understanding the world and producing meaning since the earliest times.

What is the role of technology in the production of meaning and experience? The theory of the extended mind (Clark and Chalmers 1998; Clark 2003, 2008) shows us that technologies can function as extensions of the mind and, therefore, a constitutive part of the production of meaning. Cognitive technologies, as Clark (2003, 2014, chapter 8) argues, make us natural-born cyborgs, in the sense that we have always used technologies to augment our cognitive capabilities. Malafouris (2013) also notes that we have used devices, such as tools, to help with calculation, for example, since the Babylonian culture (around 7000–3000 BC) and that the mind is, at least partially, shaped by things outside the body. The extended mind thesis, by Clark and Chalmers (1998), advances the idea that cognitive technologies can serve as constitutive parts of the human mind, in specific situations and uses. The material engagement theory of Malafouris shows how considering the world of things—artifacts, material signs—as constitutive of cognition challenges consecrated ideas about human cognitive evolution and capacities.

Take a contemporary example such as word processor programme in a computer. This technology allows you to alter texts easily, substitute texts and position images, etc., without having to erase letters on the page, as in the days of typewriters, or even to rewrite the entire page in long hand. More than this, the process of writing—however messy it may seem—helps with thinking, trying out an idea here and there in the composition of the text. It is not uncommon for people to say that such technologies aid in the writing process. Artists use cognitive tools in a similar way. A painter makes sketches of paintings, or parts of them, before executing the final painting. Not only do they experiment with colors, etc., but they create the painting itself with these sketches. In a certain sense, we can understand the materials and instruments used by the artist as cognitive technologies. With these examples, we can see that the use of technology is not foreign to art or to human cognition. In a slogan dear to Enactivist theories of mind: "we create a path in walking." The written text or the painting is not there, in the head, before we start executing it on paper or some other medium: it is enacted in the process of experimenting with artifacts and mobilizing our emotions.

But here we have to be careful. The extended mind hypothesis could include artificial systems within its scope and consider them, at least in principle, as intelligent agents capable even of emotions. However, Colombetti (2017, p. 5) notices that the extended mind hypothesis rests on a computational-functionalist view of cognition, while the Enactivist conception "offers a very different account of cognition—one that rejects computational functionalism in favour of a view of the mind based on a specific account of the organization of living systems."

In order to understand this latter conception, we have to understand some of the concepts employed in elaborating it. Specifically, it is important to deal with how the organism individualizes itself, that is, forms itself as an autonomous organism, separated in its organization and characterized by operational closure. These concepts of 'organization' and 'operational closure' come from the theory of autopoiesis, by Maturana and Varela (1980). For an organism to maintain its identity (individuation), it needs to maintain its internal organization. This organization is of structures and processes that are related in a way that forms a network of reciprocal relationships with each other. It is this network that is operationally closed. If it ruptures, the organism dies. However, this does not mean that it is separate from the environment in which it is living. On the contrary. In addition to the self-distinction involved in individuation, there is the need for self-production, that is, adaptation to the environment, to survive. In other words, although the organism is operationally closed, it is thermodynamically open; the exchange of energy and matter between it and the environment is necessary to keep it alive. In short, there is a tension between self-distinction and self-production, which is constant throughout a person's life (Di Paulo 2018).

Can this analysis be applied to artificial minds? Here, I think, we need to introduce two more concepts developed by Di Paulo and colleagues. Within the Enactivist paradigm, the concept of sense-making is "the basic mark of the cognitive" (Thompson 2011: 211). Di Paulo and colleagues (Di Paulo et al. 2018, p. 33) define sense-making in the following manner.

> Sense-making is the capacity of an autonomous system to adaptively regulate its operation and its relation to the environment depending on the virtual consequences of its own viability as a form of life. Being a sense-maker

implies an ongoing (often imperfect and variable) tuning to the world and readiness for action.

This definition reminds us of some of the central ideas of Dewey on the creation of meaning through interaction with the environment. Colombetti (2017, p. 6) observes that "This account thus departs significantly from the one provided by approaches in cognitive science and analytic philosophy of mind that characterize cognition as a computational process operating over representational items." As we have seen, it is exactly this last characterization of cognition that is assumed in artificial "cognitive" systems.

Another central concept here is that of agency. Di Paulo and colleagues (Di Paulo et al. 2017, p. 127) say the following.

> A live organism—through its dependence on interactions with the environment for its own self-individuation—has a world of significance to which it is sensitive and in which it acts. In order to sustain its precarious hold on itself, this entity must turn outward and engage this world. These engagements or behaviors are part of what constitute the agent as a whole, not just something they do apart from their being agents.

An agent, therefore, "is defined as an autonomous system capable of adaptively regulating its coupling with the environment according to norms established by its own viability conditions" (Di Paulo et al. 2017, p. 127). It is my contention that artificial systems cannot be autonomous agents in this sense; they are not live organisms. They are not sense-makers, nor do they regulate their interaction with the environment as agents in the full sense of this term.

Of course, in very simple organisms this sensitivity involves no consciousness, not even pre-reflective consciousness. However, as the organism becomes more complex and its interactions with the world more complex, the sensitivity to the environment is expressed in feelings at the pre-reflective level of interaction. Alternatively, we could refer to this as sentience, an essential capacity for sense-making. In even more complex animals, sensitivity to the environment is also expressed in emotions and, in humans, also in language; the complexity of sensitivity is determined by the complexity of the interactions involved. In human beings, it is manifested in a complex and comprehensive repertoire of feelings, emotions, sensorimotor acts and linguistic practices (Di Paulo et al. 2018).

Here, a critic could raise the same objection encountered before: what prevents artificial systems from exhibiting the characteristics of being an autonomous, operationally closed system that responds to the world through affectivity? Why can't it be considered a "live organism" as defined by this theory? One possible response to this is to say, together with Colombetti (2010, p. 146), that "emotion must be conceptualized as a faculty of the entire embodied and situated organism. Evaluations arise in this organism by virtue of its embodied and situated character, and every situated organism carries meaning as such—not through some separate abstract cognitive-evaluative faculty." In other words, "there is no room for cognition without emotion" (Colombetti 2010, p. 151). If the critic continues to argue that machines can also have emotions, I think we have to contest the computational-functionalist assumptions on which such a statement rests. My own view is that no artificial system can have or exhibit emotions because no such system can have subjective experiences, be an agent or make sense of the world, in the manner analysed by Enactivist theories of mind.

10   Artificial Art?

But what about the "art" being produced today by AI image generators? Recently, a work produced by such a system won an award in the USA. Also, there is a lot of excitement these days about the possibility of AI writing texts, as can be seen from the explosion of interest in large language models such as ChatGPT.

Now, someone might ask: How is this possible if Johnson's analysis, set out above, is correct? Doesn't this analysis imply that only living organisms, with aesthetic sensibility, feelings and emotions, can produce art? If AI systems can produce art, does this mean that aesthetic sensibility is not necessary for the production of meaning in art and, by extension, the production of any meaning whatsoever? My answer to that question is the following: AI cannot produce meaning precisely because it is incapable of aesthetic sensibility and is not a living organism. It simply copies and does not produce real, authentic art. I'll finish my discussion by elaborating a little more on this.

It is known that AI systems need to be fed by huge quantities of data to generate results. The latest ones, such as ChatGPT, need Big Data on a gigantic scale in order to display the performance that leaves so many people captivated. However, although humans need some experience and practice to produce something reasonable, the time and quantity required is much less than an artificial system. This suggests that it is not the amount of data or experience that is important but the human capacity to sensitively engage with the world. History is full of individuals who displayed exceptional talent when they were still children, with little experience. Mozart is one of the most striking examples of this.

A central dimension to authentic art is originality and imagination, with the aim of exploring something new or "portraying" something to us that has not been perceived before. No system that simply manipulates what already exists can contribute in this way, although, of course, it can produce a passable copy that, at times, could be confused with the original. Also, no artificial system would use tools and other material things to develop its art, as cognitive technologies in the way described above. Artificial systems only combine things already known and stored in data by using inductive, probabilistic procedures. They produce no new human meaning. This point requires more detailed elaboration.

In his book *The Language Animal*, Charles Taylor (2016) argues that language, in the broad sense that is not restricted only to verbal language, but includes gestures, embodied schemes, etc., allows human beings to produce human meanings, that is, meanings that create ways of being and values necessary for human flourishing. He calls these meanings "human senses" and argues that they are totally different from the meanings produced by rules or algorithms. If we understand this, we can understand why artificial systems cannot produce "human meanings."

Taylor distinguishes between two functions of language: describing the world and constituting the world. It is the constitutive purpose of language that allows new experiences and new ways of being. In this function, language constitutes our world, in the sense of a world that matters to us, of our involvements, what is important to our lives. This world exists before our birth and we are part of it. We do not create it *ex nihilo* with every use of language. However, we can transform it and create new meanings and what Taylor calls "strong evaluations" that guide our path in life. In this sense, the change that occurs is existential, the creation of a new way of being and new "human meanings."

These are meanings that the world has for us because of our bodily and affective interaction with the world, "in our relationships with [things] (…) [T]his understanding is rooted in our bodily know-how, which allows us to make our way in and around our immediate surroundings and deal with the objects that appear" (Taylor 2016, pp. 148–9). The main mechanism for this is what Taylor calls "articulation" via language, in the broad sense, which includes body language, eye contact, tone of voice and several other embodied ways of enacting a world and its meaning. Furthermore, "human senses" are always accompanied with a "felt intuition." In Taylor's (2016, p. 183) words: "what this means is that there is no dispassionate access to these meanings; that in the case of the first person, for them to be meanings for me, values that I recognize and that move me, I have to experience their felt intuition."

If something like Taylor's analysis is correct, and it is compatible with everything outlined above, we can easily see why artificial systems are incapable of producing human meaning—whether in paintings, verbal language or any other means. As I have written elsewhere (in press): obviously, an organism capable of interpretation and articulation, in the way elaborated by Taylor, is an organism that cares about its own life and the environment in which it lives. An artificial "cognitive" system, although it can perform calculations and even manipulate the logic of description—even through verbal language, as in translation systems—is not capable of the constitutive logic of language. Nothing matters for such a system; nothing affects the system, in the sense of being affected by something. Only biological cognitive systems have full linguistic capacity. Of course, such ability has levels; a simple organism can make sense at the level of bodily expression, but not through verbal language. This ability is exclusive to more advanced animals.

To sum up, artificial systems can imitate human meaning production, but they cannot produce the real thing.

Someone may complain at this point, saying: but if it can be imitated, what is the difference between the imitation and the real thing? Does it not have equal value as produced meanings? The answer to that question is "no." We can draw an analogy with art. A copy of a work of art does not have the same value as the original, because it is recognized that the original contains previously unexpressed "human meanings" and products of the artist's way of being and expressing self. The same applies to any copy, whether made by a human or an artificial system, including "works" in someone's style. As Johnson tells us, and as I mentioned earlier, an artist's work is the expression of the qualities in a situation felt by him or her and, even when copied, still belong to the artist. These are meanings produced by the artist's aesthetic sensibility.

The fact that we can be fooled by artificial systems does not mean that there is no difference between what they produce and the production of "human meanings" by human beings. Such systems are not capable of producing human meanings, even though it seems that they can. This is important, because the introduction of such systems into more and more aspects of our lives can have unintended consequences. This is because to the extent that we believe that artificial systems are producing "human meanings," we may "interact" with them in a way that reduces our own possibilities of producing such meanings. To the extent that we assume that meanings are produced by a human, we react with "felt intuitions" and our own productions of meaning, hoping for the possibility of regulating the meanings that mediate our interaction with the world. However, the artificial system cannot respond to the "human sense" embedded in our meanings. Thus, our own humanity can be diminished in the false interaction promoted by artificial systems.

11   CONCLUSION

In this essay, I have tried to present an approach to the human mind, drawing mainly on the work of John Dewey and Mark Johnson, that emphasizes the body, affectivity and aesthetic sensibility in the production of meaning. The objective was to show how human experience implies the production of human meaning, which, in turn, requires an aesthetic sensibility and an organic body. This, in turn, implies that artificial systems are not capable of experience, aesthetic sensibility and the production of human meanings.

If this analysis is correct, or at least along the right lines, we can understand why art is perhaps the paradigm of human meaning-making and why the aesthetic dimension of human experience is pervasive. We can also understand why artificial systems are not capable of producing human meaning. However, I would like to end with a word of caution: if this is true, we need to be aware of the constraints on "human meaning" production that artificial systems may create for meaning production now and in the future.

REFERENCES

Clark, A. 2003. *Natural-Born Cyborgs. Minds, Technologies, and the Future of Human Intelligence.* Oxford: Oxford University Press.
Clark, A. 2008. *Supersizing the Mind. Embodiment, Action, and Cognitive Extension.* Oxford: Oxford University Press.

Clark, A. 2014. *Mindware. An Introduction to the Philosophy of Cognitive Science.* Oxford: Oxford University Press.

Clark, A. & Chalmers, D. 1998. The Extended Mind. *Analysis* 58: 7–19.

Colombetti, G. 2010. Enaction, Sense-making, and emotion. In Stewart, J.; Gapenne, O. & Di Paolo, E. A. (Eds.) *Enaction: Towards a New Paradigm for Cognitive Science.* Cambridge, Massachusettes: MIT Press.

Colombetti, G. Enactive affectivity, extended. In *Topoi*, 36:445–455, 2017.

Dewey, J. [1896] The Reflex Arc Concept in Psychology. In *Psychological Review* 3: 357–370.

Dewey, J. 1981 [1926]. *Experience and Nature. The Later Works*, Volume 1, edited by Jo Ann Boydston. Southern Illinois University Press.

Dewey, J. 1984 [1930]. Qualitative Thought. In *The Later Works, Volume 5: 1929-1930.* Edited by Jo Ann Boydston. Southern Illinois Press.

Dewey, J. 1980 [1934]. Art as Experience. New York: The Berkeley Publishing group.

Di Paulo, E.A. 2018. The Enactive Conception of Life. In Newen, A., de Bruin, L; Gallagher, S. *The Oxford Handbook of 4E Cognition.* Oxford: Oxford University Press.

Di Paulo, E.A., Buhrmann, T., Barandiaran, X. E. 2017. *Sensorimotor Life. An Enactive proposal.* Oxford: Oxford University Press.

Di Paulo, E. A., Cuffari, E., De Jaegher, H. 2018. *Linguistic Bodies. The continuity between life and language.* Cambridge, Massachusettes: MIT Press.

Gibson, J.J. 2015. *The Ecological Approach to Visual Perception.* New York: Psychology Press.

Gumbrecht, H-U. 2003. *The Production of Presence. What Meaning cannot Convey.* Stanford: Stanford University Press.

Hutchins, E. 1995. *Cognition in the Wild.* The MIT Press.

Hutto, D.D. & Muin, E. 2013. *Radical Enactivism, Basic Minds without Content.* The MIT Press.

Johnson, M. 1987. *The Body in the Mind. The bodily basis of meaning, imagination, and reason.* Chicago: University of Chicago Press.

Johnson, M. 2007. *The Meaning of the Body. Aesthetics of Human Understanding.* Chicago: University of Chicago Press.

Johnson, M. 2017. *Embodied Mind, Meaning, and Reason. How our bodies give rise to understanding.* Chicago: University of Chicago Press.

Johnson, M. 2018. *The Aesthetics of Meaning and Thought. The bodily roots of philosophy, science, morality, and art.* University of Chicago Press.

Johnson, M, Tucker, D.M. 2021. *Out of the Cave. A natural philosophy of mind and knowing.* The MIT Press.

Johnson. M, Schulkin, J. 2023. *Mind in Nature. John Dewey, cognitive science, and a naturalistic philosophy for living.* The MIT Press.

Kurzweil, R. 2005. *The Singularity is Near.* London: Viking.

Malafouris, L. 2013. *How Things Shape the Mind. A Theory of Material Engagement.* The MIT Press.

Maturana, H., Varela, F. 1980. *Autopoiesis and Cognition: the realization of the living.* Dordrecht, The Netherlands: D. Reidel.

Merleau-Ponty, M. 1969. *The Structure of Behaviour.* Pittsburgh: Duquesne University Press.

Sheets-Johnston, M. 2011. *The Primacy of Movement.* Amsterdam: John Benjamins Publishing Group.

Taylor, C. 1989. *Sources of the Self.* Cambridge: Cambridge University Press.

Taylor, C. 2016. *The Language Animal. The full shape of the human linguistic capacity.* Cambridge: Belknap.

Thompson, E. 2011. Reply to commentaries. In *Journal of Consciousness Studies* 18(5–6):176–223.

Uexküll, J. J. Von. 2020. *A Foray Into the Worlds of Animals and Humans: With a Theory of Meaning,* translated by Joseph D. O'Neil, Minneapolis/London: University of Minnesota Press.

Index[1]

---

[1] Note: Page numbers followed by 'n' refer to notes.

© The Author(s), under exclusive license to Springer Nature
Switzerland AG 2024
P. Alexandre e Castro (ed.), *Challenges of the Technological Mind*,
New Directions in Philosophy and Cognitive Science,
https://doi.org/10.1007/978-3-031-55333-2